A Time Series Approach to Option Pricing

Christophe Chorro • Dominique Guégan •
Florian Ielpo

A Time Series Approach to Option Pricing

Models, Methods and Empirical Performances

Springer

Christophe Chorro
Dominique Guégan
Maison des Sciences Economiques
Université Paris 1 Panthéon Sorbonne
Centre d'Economie de la Sorbonne
Paris
France

Florian Ielpo
Lombard Odier Darier Hentsch & Cie
Genève
Switzerland

ISBN 978-3-662-45036-9 ISBN 978-3-662-45037-6 (eBook)
DOI 10.1007/978-3-662-45037-6
Springer Heidelberg New York Dordrecht London

Library of Congress Control Number: 2014957876

Printed on acid-free paper

Springer is part of Springer Science+Business Media (www.springer.com)

Notation

Sets and Spaces

$\emptyset$	Empty set
$\mathbb{N}$	Positive integers
$\mathbb{Z}$	Integers
$\mathbb{Q}$	Rational numbers
$\mathbb{R}$	Real numbers
$\mathbb{R}_+$	Positive real numbers
$\mathbb{R}_+^*$	Strictly positive real numbers
$[a, b]$	Closed interval
$]a, b[$	Open interval
$\{a, \ldots, b\}$	Integers between a and b
$(\Omega, \mathcal{A}, P)$	Probability space
$L^p(\Omega, \mathcal{A}, P)$	Real random variables with finite p order moment

Functions

Argmax	Argument of the maximum
Id	Identity function
e^x	Exponential of x
$log(x)$	Logarithm of x
$(x)_+$	$Max(x, 0)$
$\mathbf{1}_A(x)$	1 if $x \in A$, 0 otherwise
$\mathcal{R}e(x)$	Real part of x
K_λ	Bessel function of the third kind
N	Distribution function of a $\mathcal{N}(0, 1)$
$J(\Phi)$	Jacobian determinant of Φ

Convergence

$\rightarrow$	Pointwise convergence
$\underset{a.s}{\rightarrow}$	Convergence almost sure
$\underset{L^p}{\rightarrow}$	Convergence in $L^p(\Omega, \mathcal{A}, P)$

$\underset{\mathcal{D}}{\rightarrow}$	Convergence in distribution
limsup, liminf	limit superior, limit inferior

Probability and Finance

$E[X]$	Expectation
$E[X\|\mathcal{G}]$	Conditional expectation
$Var[X]$	Variance
$Cov[X,Y]$	Covariance
$sk[X]$	Skewness
$k[X]$	Kurtosis
γ_X	Autocovariance
ρ_X	Autocorrelation
$\frac{dQ}{dP}$	Radon–Nikodym derivative of Q w.r.t. P
$\sigma(X_u; u \leq t)$	σ-Algebra generated by $(X_u)_{u\leq t}$
$\tilde{X}_t$	Discounted value of X_t
$\mathbb{G}$	Moment generating function
$\mathbb{Q}^{em}$	Elliott–Madan equivalent martingale measure
$\mathbb{Q}^{ess}$	Esscher equivalent martingale measure
$\mathbb{Q}^{qess}$	Quadratic Esscher equivalent martingale measure

Probability Distributions

$D(0,1)$	Arbitrary distribution with mean 0 and variance 1
$G_a(a,b)$	Gamma distribution
$GH(\lambda,\alpha,\beta,\delta,\mu)$	Generalized hyperbolic distribution
$GIG(\lambda,\chi,\psi)$	Generalized inverse Gaussian distribution
$MN(\phi,\mu_1,\mu_2,\sigma_1,\sigma_2)$	Mixture of two Gaussian distributions
$\mathcal{N}(m,\sigma^2)$	Gaussian distribution with mean m and variance σ^2
$\mathcal{P}(\lambda)$	Poisson distribution

GARCH Models

APARCH	Asymmetric power ARCH
ARCH	Autoregressive conditional heteroscedasticity
EGARCH	Exponential GARCH
EWMA	Exponentially weighted moving average
GJR	Glosten, Jagannathan and Runkle GARCH
HN	Heston and Nandi model
IGARCH	Integrated GARCH

Abbreviations

AD	Andersen–Darling
AO	Arbitrage opportunity
AAO	Absence of arbitrage opportunity
AARPE	Average absolute relative pricing errors
ACF	Autocorrelation function

ACVF	Autocovariance function
ARMA	Autoregressive moving average
BFGS	Broyden–Fletcher–Goldfarb–Shanno algorithm
cadlag	Right continuous with left limits
CCAPM	Consumption capital asset pricing model
EMS	Empirical martingale simulation
FFT	Fast Fourier transform
i.i.d	Independent and identically distributed
KS	Kolmogorov–Smirnov
LRNVR	Locally risk neutral valuation relationship
MA	Moving average
ML	Maximum likelihood
NIC	News impact curve
PCA	Principal component analysis
QML	Quasi maximum likelihood
REC	Recursive estimation method
RNVR	Risk neutral valuation relationship
RMSE	Root mean squared error
SDF	Stochastic discount factor
S&P500	Standard and Poor's 500 equity index
VIX	Chicago Board Options Exchange market volatility index

Contents

List of Figures

List of Tables

Introduction

1

An equity option is a financial asset that gives its buyer the right (but not the obligation) to buy or sell a certain quantity of stocks or financial instruments on or before specified dates at a predefined price. To a certain extent, an option is similar to an equity future, except for the fact that the buyer has no commitment to buy or sell anything at the due date. Actually, options fall into two main classes (see e.g. Hull 2011): *vanilla* and *exotic* that differ in exercise styles and payoff values. As their respective names make it rather clear, vanilla options are standard financial assets with a simple type of guaranty whereas exotic ones have more complex financial structures (e.g. a Barrier option structured such that the underlying stock has to reach a certain level to be active or inactive). To avoid specific technical and numerical problems, in this book we will focus on the vanilla type as it has been used in the academic literature as a benchmark for comparing empirical performances of pricing models. Basically, we wish to help readers to understand the value that the time series methodologies presented in this book can add to this well known basis before turning to securities with a more complex payoff.

This book will focus on a certain type of vanilla options known as *European options*. This is however seldom restrictive, as European options are the most traded options around the world. A European call option with a maturity T and a strike price K, associated with a stock whose price at time $t \in [0, T]$ is denoted by S_t, is a financial contract that enables its owner to buy the stock on date T (and only on this date) at the predetermined price K. It is an option in the sense that the buyer has no legal obligation to do so, whereas the seller of the option must deliver the stock on the date T at the price K if the buyer is willing to exercise his right. When S_T is greater than K, the buyer certainly has a natural interest to exercise his option to materialize an instantaneous gain $S_T - K$, as he can sell in the market at a higher price the stock he just bought. Now, if S_T is lower than K, exercising the option would imply a loss for the buyer, as the sell price is now lower than the buy price. A call option therefore works as an insurance: it makes it possible for the buyer to know for any time t prior to T the maximum price at which he will be able to buy

C. Chorro et al., *A Time Series Approach to Option Pricing*,
DOI 10.1007/978-3-662-45037-6_1

a certain stock. As for any insurance, this comes at a cost called the *premium*. This premium is naturally proportional to the likelihood of the scenario for which S_T is greater than K. On a date t very close to T, if S_t is greater than K, the premium for the option is significantly different from zero as an immediate exercise (even though it is not possible for European options) would immediately trigger a gain for the buyer. Intuitively, the option price will be proportional to this immediate gain, offsetting it to a large extent. Now, considering t further away from T, giving a price to such an insurance contract is a trickier task as more refined intuitions are necessary. In general, it requires a mathematical modeling framework of the sort that will be derived in this book. Finally, when a call option gives the right to buy the stock at a predefined price K, a put option gives the right to symmetrically sell a stock at a preset price K, those two assets being somewhat mirrors of each other. Again, the pricing of such assets requires a suitable framework that can only be based on financial intuitions to which a layer of mathematical modeling must be added.

Back in 1973, Black and Scholes (1973) proposed a first rigorous approach to this issue.[1] The main underlying intuition is a financial one: absence of arbitrage opportunities (AAO) arguments. A financial arbitrage has nothing to do with speculation: it simply consists of building a financial portfolio which brings a certain gain immediately or in the future. To illustrate, simply look at the buyer of an option who is of course entering in such a deal only if he anticipates that he would gain a profit. Nevertheless, this gain is by nature uncertain and based on the buyer's convictions about the unknown future evolution of the stock. Therefore, this strategy leads to speculation and not to an arbitrage opportunity because it involves a significant amount of risk taking. Starting again from the option maturity date T, the payoff of a European call option is $S_T - K$ if $S_T > K$. In this case, the price of the option is exactly the final payoff. If this was not the case, a position could be built to profit from this arbitrage opportunity. If the price of the option is below $S_T - K$, then a wise investor would buy an unlimited amount of options, exercise them instantly, cashing $S_T - K$. His profit per trade would then be $S_T - K - C_T$ where C_T is the price of the option a.k.a. its premium at T. Intuitively, the seller of the option would adjust its price given this surge in demand so that $C_T = S_T - K$, canceling the arbitrage. A similar reasoning can be obtained when $C_T > S_T - K$. This arbitrage way of thinking forms the fruitful basis of option valuation in the Black and Scholes (1973) framework.

The only matter now is to be able to obtain such a reasoning for any date $t < T$: in this case, the investor would have to deal with two issues: first, he would need to guess what S_T is going to be worth at time T, somewhat forecasting it using a mathematical model and an educated guess. Second, the expected final payoff of the option is a cash flow at the maturity time T, whereas the price of the option is a time t value: the final payoff at time T needs to be converted into a time t value by

[1]The authors attribute to Robert C. Merton the key financial intuition of their seminal paper: the no-arbitrage principle.

discounting it using a proper interest rate. A fair price for this call option for a given investor would be somewhat obtained from the following formula:

$$C_t = e^{-r(T-t)} E_L \left[(S_T - K)_+ | I_t \right],$$

where I_t gathers the information available at time t about all of the relevant factors that can explain the variations of S_t until the maturity date T, r is an interest rate relevant to the investor and $(S_T - K)_+ = \max(0, S_T - K)$. The notation $E_L[.]$ indicates an expectation computed with respect to the probability L that represents the investor's forecast of S_T given the time t information set. Because of L, this is a subjective price: each investor can have a different view on the future value of S_T. However, financial assets have one unique price: or else, an arbitrage opportunity could be easily formed to profit from the difference in prices.

Black and Scholes show that in their world ruling out arbitrage opportunities, there is actually a single probability $\mathbb{Q}$ associated with a single interest rate r_f, called the risk-free rate, which makes the computation of such a price possible, turning the previous equation into the following one:

$$C_t = e^{-r_f(T-t)} E_{\mathbb{Q}} \left[(S_T - K)_+ | I_t \right].$$

Despite the fact that any investor could have his personal opinion regarding what the fair price of a specific option could be, in a market without arbitrage this price is unique and obtained from the previous fundamental equation. Again, the symmetry for insurance policies is striking: the price of an option is nothing but the value of its expected payoff brought back into today's dollars through the mechanism of actualization. The probability distribution isolated by Black and Scholes (1973) is called the *Risk Neutral* distribution because of its noteworthy property: it makes the expected return of any asset equal to the risk-free rate.

Most of the previous derivations presented in the Black and Scholes (1973) path-breaking paper are however based on very specific settings, involving two key hypotheses:

- primarily, they assume that time elapses in a continuous manner: the stock trades continuously so that at any time t there is an available transaction price S_t that allows for perfect hedging arguments. This hypothesis is essential to prevent the market from arbitrage opportunities, as the price of an asset always instantaneously reflects all of the available and relevant financial information.
- Secondarily, and probably the most important of those two hypotheses is that stocks' log-returns are assumed to be Gaussian. The Gaussian distribution is a commonly and widely used distribution that fails to cope with two essential features of financial returns: the appearance over history of extreme returns—jumps in the dynamics of returns that often reflect crises—and the asymmetry of financial stock returns—positive returns have a tendency to be of a smaller scale than negative ones. In particular, this hypothesis is essential to the derivation of

the so-called closed-form formula for the price of a European call option in the Black and Scholes pricing model.

The Black and Scholes model is therefore by itself a deep and brilliant financial idea that is based on a poor model. This could remain an academic discussion, if market participants had not fully accepted their option pricing formula. However, as illustrated in Rubinstein (1985) and in the third part of this book, the prices strictly derived from the Black and Scholes model are hardly compatible with observed market quotes. In essence, when $|S_t e^{r(T-t)} - K|$ is large—that is for strike prices which are far away from the capitalized value of S_t between t and T—the Black and Scholes prices are below market prices. This simply comes from the fact that with its underlying Gaussian distribution the model fails at mimicking extreme returns that can be observed in financial returns: there is a price for this extra uncertainty observed in financial markets that needs to be incorporated in option prices. This premium has been found to be even more significant since the 1987 market crash: Bates (1991) diagnoses the development of a crashophobia among investors leading to increased option premia for extreme strikes.

Recognizing the importance of the departure between the prices predicted by the Black and Scholes model and the option market prices obtained from the free confrontation of supply and demand, Heston (1993) was among the first to propose an alternative to the Gaussian distribution. His model eliminates the hypothesis that stock returns' volatility is constant, allowing in particular for arbitrary correlation between volatility and asset returns, and it generates prices much closer to the observed ones. However, Heston model somewhat manages to preserve one of the key features of the Black and Scholes model back in the 1990s: the closed-form expression for the premium of European vanilla options. The computing power of modern computers played an important role in the development of alternative option pricing models. Building on Heston's (1993) work, Bates's (1996) model adds jumps to Heston's (1993) stochastic volatility dynamics, bringing the prices obtained from his model closer to market quotes. Those continuous time refinements of the Black and Scholes model were finally discovered to belong to the same family of models, a family defined by their characteristic function that is an exponential affine function of their state variables as presented in Duffie and Kan (1996). Carr and Madan (1999) then showed how to use the powerful fast Fourier transform (FFT) (see Walker 1996) to price options efficiently in this framework, making of the Heston-Bates model the new benchmark of the option pricing industry.

Users of these continuous time stochastic volatility models are still faced with one significant practical issue: the estimation of the parameters of these models is extremely complex, requiring the use of highly sophisticated econometric methodologies such as state-space filters. Presumably, the complexity of these methodologies drove the industry to perform *calibration* instead of estimation. When calibrating an option pricing model, its user looks for parameter values that make the model implied prices as close as possible to observed prices using directly the risk-neutral dynamics. Estimating such a model means relying solely on the historical time series dynamics of returns to find such parameters, then turning the time series distribution into a risk neutral one in order to obtain option prices. Calibration is a very useful numerical exercise to perform what is known as option book hedging: when selling an option, the seller usually wants to hedge his position, so that when the underlying stock or any other relevant risk factor moves, it does not impact the seller of the option. By hedging his position, the option seller makes sure to preserve the premium obtained from selling the option. Calibration will then have to be performed every day until the maturity of the option, as the option model's parameters are inherently subject to daily changes: those models lacking an actual economic structure essentially because of the way parameters' values are selected. Yesterday's hedges are therefore not optimal today, and the seller's hedges must be adapted accordingly.

Time series analysis made its way to option pricing only recently. By the essence, time series analysis uses historical datasets of returns and raises the question of the best model to be used to mimic stocks' dynamics. A now well-known battery of econometric tests are available within this framework to help their users obtain a model that is consistent with the empirical distribution of returns. Relying on historical returns has *four advantages*: first, it makes it possible to build a structural model, that is tailored to match the historical properties of returns, instead of assuming what a reasonable model should be. Second, from the fact that this type of models is firstly designed based on the historical dynamics of returns, it has a more structural bent that can help reducing the frequency of computations to estimate hedging strategies as performed in Badescu et al. (2014). Third, when pricing options that are not quoted in any market, an approach based on the historical dynamics of returns contains all the necessary information to compute option prices. Calibrating a continuous time model requires an existing vanilla options market from which information regarding the model's parameters can be extracted. Fourth, time series models—at least those presented in this book—are easily estimated and many computer routines already exist to perform such an estimation. Finally, time series models are naturally designed into a discrete time framework that matches the format of returns' datasets: a list of usually equally spaced in time prices or returns on a given stock. Continuous time is a less accurate description of what actual datasets look like. Globally speaking, option pricing based on time series proposes a framework that somewhat matches a large spectrum of the requirements of the option pricing industry.

Its weak success so far can sound as a surprise to the reader given the advantages listed in the previous paragraph. However, it remains easily explainable by three major factors: first, the industry requires to be able to provide potential option buyers with a price in a couple of minutes. Up to now, the computation power of computers led a large number of participants in the pricing industry to favor closed or almost-closed-form pricing formulas: from a set of parameter values, they would be able to compute a price for an option contract—not to mention the so-called Greek letters, the optimal hedging levels that matter so much to option sellers—in only a few seconds. This limitation somewhat retreated over the past decade and Monte Carlo techniques can now be used in a very timely manner, weakening the comparative advantage of closed-form option pricing models. Then, a large number of quantitative analysts receive an intensive training in continuous time finance and have usually a more limited knowledge in statistics and econometrics, the two pillars of the discrete time option pricing methodologies presented along the lines of this book. Finally, a discrete time approach to option pricing requires to dealing with a certain number of theoretical challenges. One of them is related to the intrinsic incompleteness of markets that arises in such a framework, making it necessary to impose new assumptions regarding the shape of risk appetite.

Over the past decade, a large academic effort has been conducted to deal with the challenges raised by a time series analysis approach to option pricing. This framework should not be thought of as a competitor to the continuous time one: it is mainly grounded on the premises built by continuous time models as arbitrage theory. For example—more will be discussed along the lines of Chaps. 3 and 4—in both frameworks, the option premium is computed as the present value of the expectation of the option payoff under a risk neutral probability. Duan (1995) presented one of the first theoretical attempts to create an option pricing model based on GARCH processes. GARCH models, introduced by Engle (1982) and Engle and Bollerslev (1986), have emerged as one of the most popular discrete time alternatives to continuous time diffusions: they are time series models describing the dynamics of a stationary random sequence for which conditional variance is time varying. More precisely, Engle (1982) proposed an endogenous parametric specification of the conditional variance as a linear function of the squared past returns: therefore, a dataset containing only time series of returns makes it possible to directly estimate the joint dynamics of returns *and* volatility. In Heston (1993) model, variance is time varying as well, but it is driven by a stochastic differential equation that involves its own source of randomness: observing returns is not enough to estimate the parameters of the model as additional information about volatility is required. Duan (1995) uses a standard GARCH process together with a set of three assumptions, called the *Locally Risk Neutral Valuation Relationship* (LRNVR), directly derived from the Black and Scholes (1973) model to describe new risk neutral dynamics. First, the conditional distribution of log-returns under both historical and risk neutral probabilities is Gaussian. Then, the conditional volatility under the historical and the risk neutral probabilities is the same. Finally, under the risk neutral probability, the stock's expected return is the risk-free rate. Duan (1995) then manages to fully derive the specification of the GARCH model under a risk neutral probability fulfilling the three preceding hypotheses, showing how the model outperforms the Black and Scholes (1973) benchmark in terms of pricing performances.

Even though Duan's (1995) model does not outperform Heston's (1993) and Bates's (1996) ones, it still proved that a time series approach to option pricing can be a *possibility* available to quantitative analysts all around the world. At this moment, the academia was primarily interested in replicating one of the key features of the continuous time models seen previously: the fact that the characteristic function of the log-returns under the risk neutral distribution had a closed-form expression. Significant contributions such as Heston and Nandi (2000), Christoffersen et al. (2006) and Mercuri (2008) presented variations over this

GARCH theme using more and more sophisticated GARCH specifications, now delivering pricing performances comparable to that of the best continuous time models. A time series path to option pricing then started to be more and more accepted within the academia and an increasing number of publications started being submitted to scientific journals, unveiling and dealing with the new challenges born from discrete time models. When using a time series model, one of the main issue is probably to provide a rigorous methodology that makes it possible to move from the historical to the risk neutral dynamics. In the conditionally Gaussian case—as in Duan (1995) or Heston and Nandi (2000)—the change in probability is fully characterized by assumptions on the first two conditional moments of the distribution. Now, when moving away from such models, a more in-depth understanding of what this change in probability measure actually means is necessary. Aît-Sahalia and Lo (1998, 2000) and Jackwerth (2000) among others highlighted the fact that this change in probability actually reflects assumptions regarding the shape of investors' risk aversion. The ratio of the risk-neutral to the historical distribution is actually a function of the Arrow-Pratt (see for instance Pratt 1964) risk aversion coefficient that measures how individuals behave when being faced with situations with uncertain outcomes.

An accurate enough time series model and realistic risk aversion function were finally found to be the two key ingredients from which an option pricing model can be derived. The Black and Scholes (1973) model then became a special case of this broader way of considering option pricing: it simply is an option pricing model based on a Gaussian distribution with constant volatility (the distribution part) and a constant relative risk aversion function (the risk aversion part). A general framework to option pricing then started to be available, as illustrated by Gouriéroux and Monfort (2007). The availability of such a framework and the increasing performances of computers finally eliminated the last restraints that prevented building of a genuine time series approach, one that would not try to mimic at any cost continuous time models features such as a closed-form formula for the conditional characteristic function. This approach accomplishes the following two things: one, it looks for a model that accurately describes the historical behavior of returns; two, it considers a realistic risk aversion function. Combining the two of them makes it possible to simulate returns under the risk-neutral distribution and therefore obtain European option prices. Contributions such as Barone-Adesi et al. (2008) and Chorro et al. (2010, 2012) illustrate the challenging performances obtained, even challenging the performances obtained from the calibration of continuous time models.

This book is designed to reflect most of those recent evolutions and making it possible for both practitioners and academics to get a grasp of what time series can actually do in terms of option pricing. Its aim is not to review exhaustively all of the recent scientific achievements of the GARCH option pricing theory, but to point out in a structured and comprehensive way the main theoretical and empirical issues. The book is relatively self contained, providing interested readers with all the necessary material to be able to understand and be able to use time series analysis to price European options. It is organized as follows. Chapter 2 presents the key tools of time series analysis: we review GARCH-based models detailing both their underlying rationale and their mathematical foundations. We also discuss the estimation methodologies that come naturally with time series models and the links that may exist with diffusion processes. Chapter 3 provides a fully-fledged discussion about the transition from historical to risk-neutral probability in order to obtain option prices especially when distributions are not Gaussian. We present in details the three main approaches of option pricing in the GARCH setting namely the extended Girsanov principle of Elliott and Madan (1998), the conditional Esscher transform first used in this framework by Siu et al. (2004) and the variance dependent pricing kernel introduced in Monfort and Pegoraro (2012) and Christoffersen et al. (2013). Finally, Chap. 4 shows how to use the tools presented in Chaps. 2 and 3 using a dataset of returns on stocks and of options. At every step, R codes are presented so that the reader is able to replicate the empirical applications presented along the book. It finally provides an assessment of the pricing performances that can be obtained within this framework. Overall, the book aims at providing any interested reader with all the necessary material to use time series analysis to perform option pricing: an understanding of the economic rationale, a mastering of the elementary mathematical foundations and the associated necessary R code.

References

Aït-Sahalia Y, Lo A (1998) Nonparametric estimation of state-price densities implicit in financial asset prices. J Financ 53:499–547

Aït-Sahalia Y, Lo A (2000) Nonparametric risk management and implied risk aversion. J Econ 94:9–51

Badescu A, Elliott RJ, Ortega JP (2014) Quadratic hedging schemes for non-Gaussian GARCH models. J Econ Dyn Control 42:13–32

Barone-Adesi G, Engle RF, Mancini LA (2008) GARCH option pricing model in incomplete markets. Rev Financ Stud 21(3):1223–1258

Bates DS (1991) The crash of '87 was it expected? The evidence from options markets. J Financ 46:1009–1044

Bates DS (1996) Jumps and stochastic volatility: exchange rate processes implicit in deutsche mark options. Rev Financ Stud 9(1):69–107

Black F, Scholes M (1973) The pricing of options and corporate liabilities. J Polit Econ 81:637–659

Carr P, Madan D (1999) Option valuation using the fast Fourier transform. J Comput Financ 2(4):61–73

Chorro C, Guégan D, Ielpo F (2010) Martingalized historical approach for option pricing. Financ Res Lett 7(1):24–28

Chorro C, Guégan D, Ielpo F (2012) Option pricing for GARCH type models with generalized hyperbolic innovations. Quant Financ 12(7):1079–1094

Christoffersen P, Heston SL, Jacobs K (2006) Option valuation with conditional skewness. J Econ 131:253–284

Christoffersen P, Heston SL, Jacobs K (2013) Capturing option anomalies with a variance-dependent pricing kernel. Rev Financ Stud 26(8):1963–2006

Duan JC (1995) The GARCH option pricing model. Math Financ 5:13–32

Duffie D, Kan R (1996) A yield-factor model of interest rates. Math Financ 6(4):379–406

Elliott R, Madan D (1998) A discrete time equivalent martingale measure. Math Financ 2(8):127–152

Engle RF (1982) Autoregressive conditional heteroscedasticity with estimates of the variance of United Kingdom inflation. Econometrica 50:987–1007

Engle RF, Bollerslev T (1986) Modelling the persistence of conditional variances. Econ Rev 94:405–420

Gouriéroux C, Monfort A (2007) Econometric specification of stochastic discount factor models. J Econ 136(2):509–530

Heston SL (1993) A closed-form solution for options with stochastic volatility with applications to bond and currency options. Rev Financ Stud 6(2):327–343

Heston SL, Nandi S (2000) A closed-form GARCH option valuation. Rev Financ Stud 13:585–625

Hull JC (2011) Options, futures and other derivatives, 8th edn. Prentice Hall, Boston

Jackwerth J (2000) Recovering risk aversion from option prices and realized returns. Rev Financ Stud 13:433–451

Mercuri L (2008) Option pricing in a garch model with tempered stable innovations. Financ Res Lett 5(3):172–182.

Monfort A, Pegoraro F (2012) Asset pricing with second-order esscher transforms. J Bank Financ 36(6):1678–1687

Pratt JW (1964) Risk aversion in the small and in the large. Econometrica 32:122–136

Rubinstein M (1985) Nonparametric tests of alternative option pricing models using all reported trades and quotes on the 30 most active CBOE option classes from August 23, 1976 through August 31, 1978. J Financ 40:455–480

Siu TK, Tong H, Yang H (2004) On pricing derivatives under GARCH models: a dynamic Gerber-Shiu approach. N Am Actuarial J 8:17–31

Walker JS (1996) Fast Fourier transforms, 2nd edn. CRC, Boca Raton

The Time Series Toolbox for Financial Returns 2

The evaluation of financial risks and the pricing of financial derivatives are based on statistical models trying to encompass the main features of underlying asset prices. From the seminal works of Bachelier (1900) based on Gaussian distributions, the random walk hypothesis for the returns or the log-returns has frequently been suggested. Its remarkable mathematical tractability, in particular in the multidimensional case, was the keystone of nice financial theories like Markowitz (1959)'s portfolio management or Black and Scholes (1973) option pricing model, among others. Nevertheless, during the last decades, the explosion of computational tools efficiency has allowed researchers to pay more attention to the analysis of financial datasets and the test of models assumptions. It is now well-documented that in spite of their huge heterogeneity concerning the nature of financial assets (stocks, commodities, interest rates, currencies...), the frequency of observations or the multiplication of financial centers, financial time series exhibit common statistical regularities (called stylized facts) that make satisfactory models difficult to obtain. A major attempt in this direction was done during the 1980s by Engle (1982) and Bollerslev (1986) through the ARCH/GARCH approach. After a brief reminder of the classical stylized facts observed for the daily log-returns of financial indices, the aim of the chapter is to present the main features of the GARCH modelling approach and its recent extensions.

2.1 Stylized Facts

Mandelbrot (1963) and Fama (1965) were the first to empirically question the Gaussian random walk hypothesis bringing to light various statistical properties of asset returns. Their studies paved the way to intensive empirical works trying to exhibit statistical regularities common across a wide range of financial datasets: the stylized facts (see for example Cont 2001; Terasvirta and Zhao 2011). In this section

C. Chorro et al., *A Time Series Approach to Option Pricing*,
DOI 10.1007/978-3-662-45037-6_2

we briefly present the main stylized facts of daily log-returns of financial series [see Dacorogna et al. (2001) for an introduction to high-frequency datasets] namely:

- absence of autocorrelation for price variations, autocorrelation of the squared log-returns;
- conditional heteroscedasticity, volatility clustering, leverage effects;
- fat-tailed and asymmetric distributions.

We refer the reader to Taylor (1986) and Poon (2005) for more details on this topic. As an illustration, we include an elementary empirical study of the S&P500 index closing prices from January 3, 1990 to April 18, 2012.

2.1.1 Stationarity, Ergodicity and Autocorrelation

We consider a discrete time stochastic process $(X_t)_{t \in \mathbb{N}}$ defined on a probability space $(\Omega, \mathcal{A}, P)$ with real values. Even if it is well-known in finance that past values do not perfectly reflect the future, the requirement of any basic statistical analysis is a kind of temporal stability to be able to describe properties independent of the observation time. The following definition introduces two notions of stationarity that may replace in a natural way the hypothesis of independent and identically distributed (i.i.d) observations that is too restrictive in time series analysis.

Definition 2.1.1

(a) The process $(X_t)_{t \in \mathbb{N}}$ is strictly stationary if $\forall t \in \mathbb{N}$ and $\forall h \in \mathbb{N}$, the distributions of the random vectors $(X_0, \ldots, X_t)$ and $(X_h, \ldots, X_{t+h})$ are the same.

(b) The process is second order stationary if:
 (i) $\forall t \in \mathbb{N}$, $E[X_t^2] < \infty$;
 (ii) $\forall t \in \mathbb{N}$, $E[X_t]$ is constant;
 (iii) $\forall t \in \mathbb{N}$, $\forall h \in \mathbb{N}$, $Cov[X_t, X_{t+h}]$ is independent of t.

The functions $\gamma_X(h) = Cov[X_t, X_{t+h}]$ and $\rho_X(h) = \frac{\gamma_X(h)}{\gamma_X(0)}$ are respectively called the autocovariance (ACVF) and autocorrelation (ACF) functions of the process $(X_t)_{t \in \mathbb{N}}$.

Remark 2.1.1

(a) A stationary process fulfilling $E[X_0^2] < \infty$ is second order stationary.

(b) A second order stationary process with $E[X_0] = 0$ and $\gamma_X(h) = 0$ $\forall h \in \mathbb{N}^*$ is called a weak white noise. It is important to remark that the zero autocorrelation hypothesis does not imply in general the independence of the terms of the sequence.

(c) A sequence $(X_t)_{t\in\mathbb{N}}$ of i.i.d random variables is a stationary process that is a particular example of a weak white noise when it is centered with finite moments of order two. In this case we say that $(X_t)_{t\in\mathbb{N}}$ is a strong white noise.

Now we present the sample counterparts of the ACFV and ACF functions. When $(X_t)_{t\in\mathbb{N}}$ is a second order stationary process we define $\forall h \in \mathbb{N}, \forall n > h$,

$$\gamma_{n,X}(h) = \frac{1}{n}\sum_{i=1}^{n-h}(X_i - \bar{X}_n)(X_{i+h} - \bar{X}_n) \text{ and } \rho_{n,X}(h) = \frac{\gamma_{n,X}(h)}{\gamma_{n,X}(0)}$$

where $\bar{X}_n = \frac{1}{n}\sum_{i=1}^{n} X_i$ is the empirical mean of order n. When the X_i are i.i.d random variables, the Strong Law of Large Numbers ensures the consistency of the preceding estimators and confidence intervals may be classically obtained from the Central Limit Theorem if moments of order four are finite (Theorems A.1 and A.2 of the Appendix). Unfortunately, as we will see below, the i.i.d hypothesis for financial time series is far from reality. In the following we will say for simplicity that the stochastic process $(X_t)_{t\in\mathbb{N}}$ is ergodic if $\gamma_{n,X}(h)$ and $\rho_{n,X}(h)$ are consistent or more generally if all the estimators of moments are consistent when they exist. We refer the reader to Brockwell and Davis (1996) for a strong mathematical definition of ergodicity and for a complete analysis of the covariance structure of sample autocovariance and autocorrelation estimators when the i.i.d hypothesis is removed. As remarked in Davis and Mikosch (1998), conclusions coming from an ACF analysis of log-returns powers have to be interpreted carefully even if this is a first interesting empirical step. In fact, financial datasets may generate large confidence bands especially in the presence of heavy-tails.

Let (S_t) denote the price of a financial asset at time t. As we can see in Fig. 2.1 where the daily closing prices of the S&P500 from January 3, 1990 to April 18, 2012 are plotted, the sample paths of financial prices are generally close to a random walk

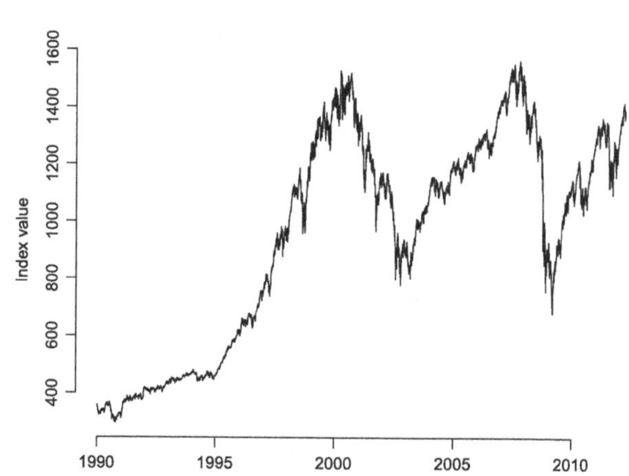

Fig. 2.1 S&P500 closing prices from January 3, 1990 to April 18, 2012 (5,619 points)

behavior and so are incompatible with the second order stationarity hypothesis due to the presence of upward and downward trends.

Nevertheless, taking the log-difference of two consecutive prices we obtain the log-returns sequence

$$Y_t = log(S_t) - log(S_{t-1}).$$

This quantity, that is in general equivalent to the relative returns $\frac{S_t - S_{t-1}}{S_{t-1}}$, has some particularly interesting features. First, from a technical point of view, time aggregation is simpler than using relative returns: returns over sub-periods are simple functions of single day returns by the additivity property. Then, as we can see in Fig. 2.2, this sequence seems to be constant around zero in average in spite of great fluctuations. In Fig. 2.3 we represent the sample autocorrelations of (Y_t)

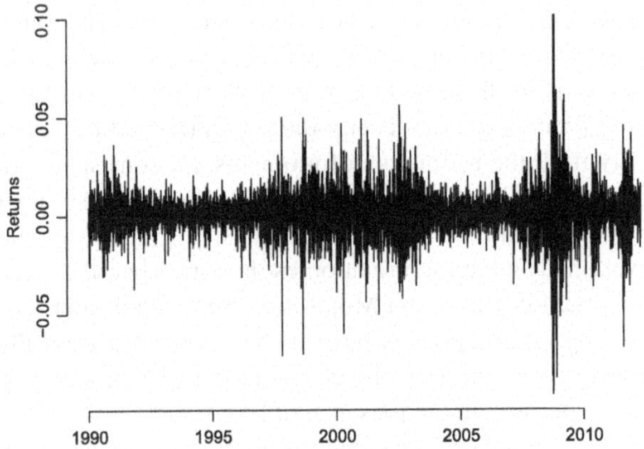

Fig. 2.2 S&P500 daily log-returns from January 3, 1990 to April 18, 2012

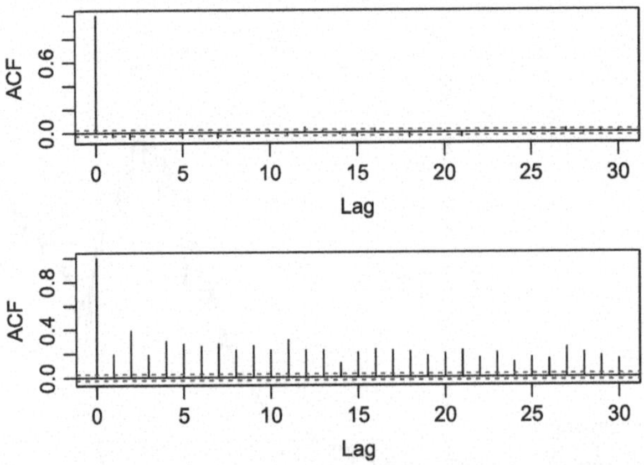

Fig. 2.3 Sample autocorrelations of the S&P500 daily log-returns (*top*) and squared log-returns (*bottom*) from January 3, 1990 to April 18, 2012

and (Y_t^2). The log-returns show no evidence of serial correlation, but the squared log-returns are positively autocorrelated. Thus a white noise without independence hypotheses (the weak one) seems to be compatible with the log-returns dynamics.

2.1.2 Time-Varying Volatility and Leverage Effects

The reactivity of the log-returns to economic and political events seems to be hardly compatible with the seminal Black and Scholes assumption of a constant (conditional) variance. The historical volatility (also simply called the volatility) is defined as the conditional standard deviation of the daily log-returns given past events:

$$h_t = (Var[Y_t \mid Y_{t-1}, \ldots, Y_1])^{\frac{1}{2}}.$$

The volatility of financial assets is a key feature for measuring the time-varying risk perception underlying investment decisions. It reflects in general the cyclical global economic stance of the world matching the National Bureau of Economic Research dating of economic crises. Without any model assumption, its statistical estimation is known to be difficult and high-frequency data are generally employed. The recent financial econometrics literature has devoted much attention to its computation and we refer the reader to Andersen et al. (2010) for a recent review. In Fig. 2.4, we have

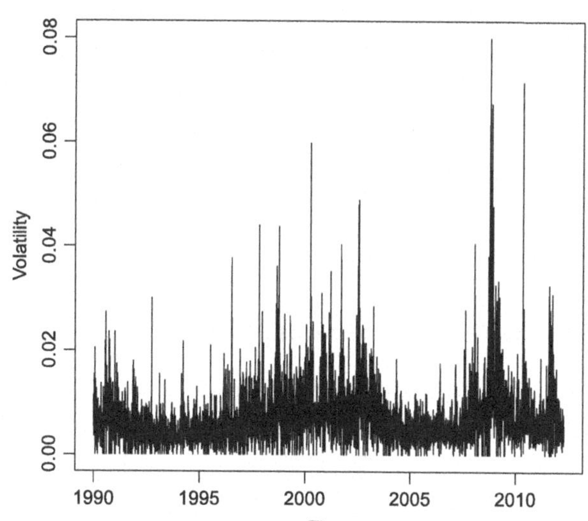

Fig. 2.4 High-low estimation of the daily volatility of the S&P500 log-returns from January 3, 1990 to April 18, 2012. At any time t, the volatility h_t is estimated by $\left(log(\frac{H_t}{S_t})log(\frac{H_t}{O_t}) + log(\frac{D_t}{S_t})log(\frac{D_t}{O_t})\right)^{\frac{1}{2}}$ where O_t, D_t, H_t and S_t are respectively the opening, the lowest, the highest and the closing prices of the S&P500 at time t

Table 2.1 Sample correlations between Y_{t-h} and Y_t^2, between Y_{t-h}^+ and Y_t^2 and between Y_{t-h}^- and Y_t^2

h	1	2	3	4	5
ρ_{Y_{t-h},Y_t^2}	−0.105	−0.113	−0.076	−0.094	−0.085
$\rho_{Y_{t-h}^+,Y_t^2}$	0.128	0.104	0.054	0.089	0.095
$\rho_{Y_{t-h}^-,Y_t^2}$	0.193	0.277	0.172	0.233	0.225

decided for simplicity to graph the volatility of the S&P500 log-returns estimated using the High-Low estimator proposed in Rogers and Satchell (1991).

As observed in Fig. 2.2 large price changes come in bulks. The volatility clustering corresponds to periods of quiescence and turbulence that tend to cluster together: when the market is shocked, it takes some time for the shock to vanish. Moreover, periods of high volatility have a tendency to last for periods of a lower length than low volatility ones (see Fig. 2.4). A quantity commonly used to measure volatility clustering is the autocorrelation function of the squared (or absolute) returns. Figure 2.3 shows that this autocorrelation function remains positive and decays slowly, remaining significantly positive over weeks. Formally, this property can be identified with what is now known as ARCH effects. This phenomenon was first observed by Mandelbrot (1963) in commodity prices. Since the pioneering papers of Engle (1982) and Bollerslev (1986) on autoregressive conditional heteroscedastic (ARCH) models and their generalization to GARCH models, volatility clustering has been shown to be present in a wide variety of financial assets including stocks, market indexes, exchange rates, interest rate securities among others.

Another well-known stylized fact is the so-called leverage effect, first discussed by Black (1976), who observed that volatility is higher during periods of negative returns and that negative returns contribute more to a rise in volatility than positive ones. In Table 2.1, we compute for the S&P500 index the correlations between Y_{t-h}, $Y_{t-h}^+ = Max(Y_{t-h}, 0)$ (positive shock), $Y_{t-h}^- = Max(-Y_{t-h}, 0)$ (negative shock) and Y_t^2 and we confirm this classical empirical fact remarking that the correlation between Y_{t-h} and Y_t^2 is negative and that the correlation between Y_{t-h}^+ and Y_t^2 is less than between Y_{t-h}^- and Y_t^2.

2.1.3 Semi Fat-Tailed Distributions

Elementary descriptive statistics for the S&P500 log-returns are provided in Table 2.2. In particular the sample skewness and kurtosis are far form their Gaussian counterparts.[1]

The distribution of the log-returns is slightly negatively skewed and the kurtosis value suggests tails that are heavier than the tails of a Gaussian distribution (in

[1]When they exist, the skewness and the kurtosis of a random variable X are defined by $sk[X] = \frac{E[(X-E[X])^3]}{Var(X)^{\frac{3}{2}}}$ and $k[X] = \frac{E[(X-E[X])^4]}{Var(X)^2}$. For a Gaussian random variable they are respectively equal to 0 and 3. These indexes are commonly used to quantify asymmetry and fat tails of distributions.

Table 2.2 Descriptive statistics of the S&P500 log-returns from January 3, 1990 to April 18, 2012

Number of observations	Mean	Minimum	Maximum
5,619	0.00024	−0.09469	0.10957

Median	Annualized standard deviation	Skewness	Kurtosis
0.00055	0.18672	−1.03129	30.688

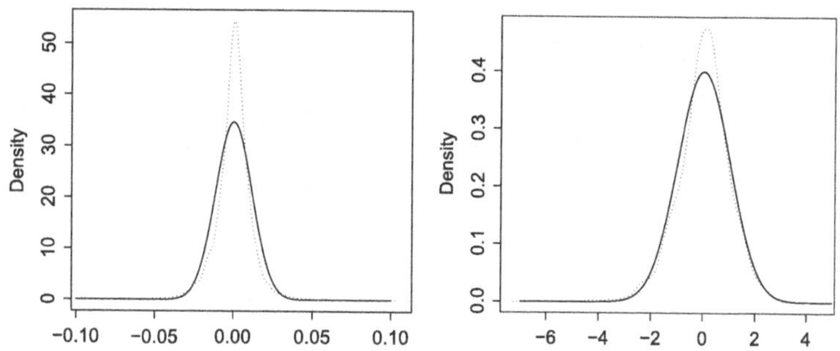

Fig. 2.5 Gaussian kernel estimators (*dotted line*) of the densities of the S&P500 daily log-returns (*left*) and of the associated residuals (*right*) from January 3, 1990 to April 18, 2012. These estimators are compared to Gaussian densities (*solid line*) with means and variances been equal to the corresponding sample means and variances. Time-varying variance has been filtered out using a GARCH(1,1) process to obtain the residuals

other words the tails decrease slower than e^{-x^2}). First pointed out by Mandelbrot (1963), the deviation from the classical Gaussian hypothesis is now well known and documented, see for example Bouchaud and Potters (2000) or Embrechts et al. (2005). In Fig. 2.5 we compare a Gaussian kernel estimator[2] of the density of the log-returns with respect to a Gaussian density with the same mean and the same variance. Clearly a sharp peak around zero appears giving more probability mass in the tails. As suggested again by Fig. 2.5, even after correcting returns for volatility effects using a GARCH(1,1) filter (see next section), the residuals still exhibit heavy tails even if tails are lighter than in the case of the unconditional distribution. Taking into account the possibility of large values of the log-returns (extreme events) is fundamental for the pricing of associated derivatives and for portfolio risk management. As we will see in Chap. 4, Gaussian hypotheses will lead in general to poor pricing performances in particular because they may usually fail to capture the short term behavior of index option smiles.

[2]When $(S_1, \ldots, S_T)$ is a sample drawn from a distribution with density f, the Gaussian kernel estimator of f of bandwidth h is given by $\hat{f}(x) = \frac{1}{Th} \sum_{i=1}^{T} d\left(\frac{x-S_i}{h}\right)$ where d is the standard normal density function. In practice we take $h = \left(\frac{4\sigma^5}{3T}\right)^{\frac{1}{5}}$ where σ is the standard deviation of the sample. This method is implemented in R via the *density* command of the *stats* package.

Even if there is no consensus on the precise quantification of the tail indexes, when decreasing sampling frequencies, log-returns classically become more and more Gaussian suggesting at least a finite moment of order 2. In fact, in the case of the log-returns, decreasing sampling frequencies is equivalent to time aggregation. Thus, the convergence of long time horizon returns toward Gaussian distribution is related to extensions of the Central Limit Theorem that need finite moment conditions. In Sect. 2.4 several semi fat-tailed distributions (allowing for extreme events and finite variances) will be suggested. From the empirical properties previously described, one can conclude that a parametric family will be able to fit the log-returns (or the associated residuals) if it has at least four parameters describing the location, the volatility, the tails behavior and the asymmetry.

Now, to overcome the imperfections surrounding Black–Scholes framework we introduce a family of discrete time models able to cope with most of the empirical features mentioned above: The GARCH models that have become, during the last decades, an important toolbox in empirical asset pricing and financial risk management.

2.2 Symmetric GARCH Models

2.2.1 From Linear to Non-linear Models

The historical importance of the linear statistical modelling is partly explained by the so-called Wold's (1938) theorem that is probably one of the most crucial result of the time series analysis. It ensures that a second order stationary process may be decomposed as an infinite moving average of a weak white noise plus a purely deterministic part:

Theorem 2.2.1 *Let $(X_t)_{t\in\mathbb{Z}}$ be a real valued stochastic process defined on a probability space $(\Omega, \mathcal{A}, \mathbb{P})$. If $(X_t)_{t\in\mathbb{Z}}$ is centered and second order stationary then*

$$X_t = \sum_{j=0}^{+\infty} b_j z_{t-j} + \eta_t \tag{2.1}$$

where $(z_t)_{t\in\mathbb{Z}}$ is a weak white noise, $b_0 = 1$, $\sum_{j=0}^{+\infty} b_j^2 < \infty$ and where $(\eta_t)_{t\in\mathbb{Z}}$ is a square integrable stochastic process such that, $\forall (t,s) \in \mathbb{Z}^2$, $E[\eta_t z_s] = 0$ and $\forall s \in \mathbb{N}$, $E[\eta_{t+s} \mid \mathcal{F}_{t-1}] = \eta_{t+s}$ with $\mathcal{F}_{t-1} = \sigma(X_u; u \leq t-1)$.

Starting from the representation (2.1), truncated formulations have been proposed to approximate purely indeterministic ($\eta_t \equiv 0$) and centered general second order stationary processes considering finite moving average of order q ($MA(q)$)

$$X_t = z_t + \sum_{j=1}^{q} b_j z_{t-j}.$$

The quality of the approximation depending on the size of q, we may add an autoregressive component of order p to the preceding equation in order to improve the precision with a parsimonious parameterization. We obtain in this case the $ARMA(p,q)$ models (see Box et al. 2008, Chap. 3, for the $MA(\infty)$ representation of an $ARMA(p,q)$ process):

$$X_t = z_t + \sum_{j=1}^{q} b_j z_{t-j} + \sum_{i=1}^{p} a_i X_{t-i}.$$

In the following example, we see why these two approaches obviously fail to reproduce the empirical volatility dynamics described in the preceding section. Let $(X_t)_{t\in\mathbb{Z}}$ be an $ARMA(1,1)$ model

$$X_t = z_t + a_1 X_{t-1} + b_1 z_{t-1}$$

where $(z_t)_{t\in\mathbb{Z}}$ is supposed to be, for simplicity, a strong white noise with $Var[z_t] = \sigma$. In this case, it is not difficult to see that

$$Var[X_t] = a_1^2 Var[X_{t-1}] + \sigma^2(b_1^2 + 2a_1b_1 + 1)$$

thus, the second order stationarity of the process is satisfied once $|a_1| < 1$ and under this condition

$$Var[X_t] = \frac{\sigma^2(b_1^2 + 2a_1b_1 + 1)}{1 - a_1^2}.$$

Nevertheless, from the classical properties of conditional expectation (see Proposition A.2 of the Appendix), $Var[X_t \mid \mathcal{F}_{t-1}] = \sigma^2$. The stochastic process is conditionally homoscedastic (constant conditional variance) a property hardly compatible with the empirical features of financial time series of log-returns.

One possible way to introduce conditional heteroscedasticity in the preceding models is to include rectangle terms as suggested in Granger and Andersen (1978). If we consider the following dynamics:

$$X_t = z_t + a_1 X_{t-1} + b_1 z_{t-1} + c_1 z_t X_{t-1}$$

the second order stationarity condition becomes $|a_1 + \sigma^2 c_1^2| < 1$ and

$$Var[X_t \mid \mathcal{F}_{t-1}] = (1 + c_1 X_{t-1})^2 \sigma^2.$$

A slight different and classical approach is to consider a multiplicative decomposition of the form

$$X_t = \sqrt{h_t} z_t \tag{2.2}$$

where

(i) the z_t are i.i.d random variables with $E[z_t] = 0$ and $Var[z_t] = 1$,
(ii) z_t is independent of $\mathcal{F}_{t-1}$,
(iii) h_t is $\mathcal{F}_{t-1}$ measurable.

Then, if (X_t) is second order stationary, we have

$$E[X_t \mid \mathcal{F}_{t-1}] = 0, \; E[X_t^2 \mid \mathcal{F}_{t-1}] = h_t$$

and $\forall h \neq 0$,

$$E[X_t] = 0, \; E[X_{t+h} X_t] = E[X_{|h|} X_0] = 0.$$

In particular (X_t) is a weak white noise and the volatility process associated to (X_t) is not constant and may reproduce the main stylized facts once h_t is well chosen. In the next section we will see that this idea is the cornerstone of the GARCH type modelling. Moreover, supposing the existence of fourth order moments for z_t and X_t we obtain

$$E[X_t^2] = E[E[X_t^2 \mid \mathcal{F}_{t-1}]] = E[h_t] E[z_t^2] = E[h_t]$$

and

$$E[X_t^4] = E[E[X_t^4 \mid \mathcal{F}_{t-1}]] = E[h_t^2] E[z_t^4]$$

from which we deduce the following expression for the kurtosis $k[X_t]$ of X_t:

$$k[X_t] = k[z_t] \frac{E[h_t^2]}{E[h_t]^2} \geq k[z_t]. \tag{2.3}$$

The equality in the preceding Cauchy–Schwarz inequality holds if and only if h_t is constant. Thus, the leptokurticity of financial time series mentioned in the preceding section may be obtained using a leptokurtic distribution for z_t or/and imposing a particular non constant form for h_t.

For the sake of completeness, let us mention that when the stochastic processes (z_t) and (h_t) are supposed to be independent (this is not the case with the previous hypotheses) we obtain the so-called stochastic volatility models. They will not be discussed in this book because they notably require specific estimation tools for the latent process (h_t) that are out of the scope of our study [we refer the reader to Taylor (1986, 2005) for a comprehensive survey].

2.2.2 Definitions

Engle (1982) was the first to consider in his seminal paper an endogenous parametric specification of the volatility process h_t in Eq. (2.2) in order to represent the time-varying conditional volatility of the monthly United Kingdom inflation through the autoregressive conditional heteroscedasticity (ARCH) model. This interesting parametric form allows the volatility to be a linear function of the past realizations of the process X_t^2. Such an expression recognizes, in particular, that the conditional forecast variance (and so the uncertainty due to this unpredictability) depends upon past information. This model, first introduced for economic purposes, proved to be extremely successful for daily financial data and we refer the reader to Bollerslev et al. (1992) and Gouriéroux (1997) for a survey of various financial applications.

Definition 2.2.1 The process $(X_t)_{t\in\mathbb{Z}}$ is an autoregressive conditional heteroscedasticity process of order $p \in \mathbb{N}^*$ (ARCH(p)) if

$$\begin{cases} X_t = \sqrt{h_t} z_t \\ h_t = a_0 + \sum_{i=1}^{p} a_i X_{t-i}^2 \end{cases} \tag{2.4}$$

where the $(z_t)_{t\in\mathbb{Z}}$ are i.i.d random variables with $E[z_t] = 0$ and $Var[z_t] = 1$ and where the a_i are nonnegative constants such that $a_0 > 0$.

From Eq. (2.4), we can see that z_t (resp. h_t) is independent of (resp. measurable with respect to) the past σ-algebra $\mathcal{F}_{t-1} = \sigma(X_u; u \leq t-1)$. When they exist, the two first conditional moments of (X_t) are given by the following relations

$$E[X_t \mid \mathcal{F}_{t-1}] = \sqrt{h_t} E[z_t \mid \mathcal{F}_{t-1}] = 0 \tag{2.5}$$

$$Var[X_t \mid \mathcal{F}_{t-1}] = E[X_t^2 \mid \mathcal{F}_{t-1}] = h_t E[z_t^2 \mid \mathcal{F}_{t-1}] = h_t = a_0 + \sum_{i=1}^{p} a_i X_{t-i}^2. \tag{2.6}$$

In particular, this time-varying volatility structure is compatible with the volatility clustering effect: there is a tendency for extreme values of the process X_t to be followed by other extreme values, but of unpredictable sign. Moreover, under integrability conditions (see Proposition 2.2.3 below) the innovation process $\nu_t = X_t^2 - h_t$ associated to (X_t^2) is a weak white noise and we deduce from

$$X_t^2 = a_0 + \sum_{i=1}^{p} a_i X_{t-i}^2 + \nu_t \tag{2.7}$$

the autoregressive representation of order p of X_t^2 that may be associated to the ARCH(p) process. This expression partly explains why for ARCH models, the decay rate of the unconditional autocorrelation function of (X_t^2) is in general too rapid compared to what is typically observed in financial time series (see Fig. 2.3), unless the maximum lag p is big. One classical alternative to obtain more parsimonious models is to add a moving average component of order q in Eq. (2.7)

$$X_t^2 = a_0 + \sum_{i=1}^{p} a_i X_{t-i}^2 + \nu_t + \sum_{j=1}^{q} \nu_{t-j}.$$

This idea, introduced by Bollerslev (1986) leads to the definition of the so-called Generalized autoregressive conditional heteroscedasticity (GARCH) model allowing for a much more flexible lag structure:

Definition 2.2.2 The process $(X_t)_{t\in\mathbb{Z}}$ is a generalized autoregressive conditional heteroscedasticity process of order $(p,q) \in (\mathbb{N}^*)^2$ (GARCH(p,q)) if

$$\begin{cases} X_t = \sqrt{h_t} z_t \\ h_t = a_0 + \sum_{i=1}^{p} a_i X_{t-i}^2 + \sum_{j=1}^{q} b_j h_{t-j} \end{cases} \tag{2.8}$$

where the $(z_t)_{t\in\mathbb{Z}}$ are i.i.d random variables with $E[z_t] = 0$ and $Var[z_t] = 1$ and where the a_i and b_j are nonnegative constants such that $a_0 > 0$.

Remark 2.2.1 In Eq. (2.8), the presence of lagged conditional variance leads to a volatility clustering that is more conspicuous for GARCH processes than for pure ARCH ones. The nonnegativity conditions on the coefficients are made to ensure the strict positivity of h_t. These conditions are necessary and sufficient for the GARCH(1,1) model but may be in general relaxed for more general structures as proved in Nelson and Cao (1992). For example, in the GARCH(2,1), less stringent necessary and sufficient conditions are given by $a_0 > 0$, $a_1 \geq 0$, $b_1 \geq 0$ and $b_1 a_1 + a_2 \geq 0$.

From now on, we focus the simplest but often very useful GARCH process, the GARCH(1,1) process given by

$$\begin{cases} X_t = \sqrt{h_t} z_t \\ h_t = a_0 + a_1 X_{t-1}^2 + b_1 h_{t-1}. \end{cases} \tag{2.9}$$

Mainly done for pedagogical reasons, this restriction is also explained by the scope of this book: the pricing of financial derivatives. As remarked in Christoffersen and Jacobs (2004), there is no empirical support for higher order lag structures especially having option pricing in mind. Moreover, the main properties of (2.9) may be explicitly deduced by means of elementary techniques without any background in advanced linear algebra. We refer the reader to Francq and Zakoian (2010), Chap. 2 for a deep theoretical study of GARCH(p,q) processes.

2.2.3 Stationarity Properties

In spite of the simplicity of equations defining (2.9), it is not clear that a strictly or second order stationary solution exists. A necessary and sufficient condition for the strict stationarity in the GARCH(1,1) case was given by Nelson (1990a) and was extended in the general GARCH(p,q) case by Bougerol and Picard (1992). For the second order stationarity, a criterion was given by Bollerslev (1986) in his seminal paper. In what follows, we remind the main lines of theses arguments in order to make the reader familiar with recursive arguments that are the key stone in the study of stochastic recurrence equations defining the volatility process in the GARCH setting. From the GARCH(1,1) specification (2.9),

$$h_t = a_0 + A(z_{t-1}) h_{t-1} \tag{2.10}$$

where $A(x) = a_1 x^2 + b_1$. Thus the process h_t admits an autoregressive representation of order one with a random coefficient. Iterating the preceding equation we obtain that, $\forall N \in \mathbb{N}$,

$$h_t = \underbrace{a_0 \sum_{j=0}^{N} \prod_{k=1}^{j} A(z_{t-k})}_{S_N(t)} + \prod_{k=1}^{N+1} A(z_{t-k})\, h_{t-N-1}, \tag{2.11}$$

where the $S_N(t)$ are the partial sums of the series with nonnegative terms

$$S_\infty(t) = a_0 \sum_{j=0}^{\infty} \prod_{k=1}^{j} A(z_{t-k}).$$

Remarking that

$$S_N(t) = a_0 + A(z_{t-1})S_{N-1}(t-1)$$

we have

$$S_\infty(t) = a_0 + a_1 X_{t-1}^2 + b_1 S_\infty(t),$$

and the existence of a strictly stationary solution to (2.9) is equivalent to the finiteness of $S_\infty(t)$. Since,

$$\prod_{k=1}^{j} A(z_{t-k})^{\frac{1}{j}} = e^{\frac{1}{j}\sum_{k=1}^{j} log(A(z_{t-k}))}$$

we obtain, from the Strong Law of Large Numbers, the almost sure convergence of $\prod_{k=1}^{j} A(z_{t-k})^{\frac{1}{j}}$ toward $e^{E[log(A(z_t))]}$. From the Cauchy criterion,[3] the series with nonnegative terms $S_\infty(t)$ is finite if $E[log(A(z_t))] < 0$. Moreover, when $E[log(A(z_t))] \geq 0$,

$$limsup \sum_{k=1}^{j} log(A(z_{t-k})) \underset{j\to+\infty}{\longrightarrow} +\infty \text{ almost surely}$$

because $\sum_{k=1}^{j} log(A(z_{t-k}))$ is a random walk with nonnegative drift. In this case, $S_\infty(t) = +\infty$. Thus, we obtain the following proposition and we refer the reader to Nelson (1990a) for the proof of the associated uniqueness:

Proposition 2.2.1 *The model* (2.9) *has a unique strictly stationary solution if and only if*

$$E[log(a_1 z_t^2 + b_1)] < 0. \tag{2.12}$$

In this case, the unique strictly stationary solution is of the form $X_t = \sqrt{h_t} z_t$ *where*

$$h_t = a_0 + a_0 \sum_{j=1}^{\infty} \prod_{i=1}^{j} (a_1 z_{t-i}^2 + b_1). \tag{2.13}$$

[3] If $(U_n)_{n\in\mathbb{N}}$ is a sequence of nonnegative terms the associated series is convergent if $limsup\, u_n^{\frac{1}{n}} < 1$.

Moreover, this solution is non anticipative in the sense that X_t *is* $\sigma(z_u; u \leq t)$ *measurable.*

Remark 2.2.2

(a) From Jensen inequality, $E[log(a_1 z_t^2 + b_1)] \leq log(E[a_1 z_t^2 + b_1]) = log(a_1 + b_1)$. Thus, condition (2.12) is fulfilled when $a_1 + b_1 < 1$.
(b) In the ARCH(1) case ($b_1 = 0$), the condition (2.12) becomes $log(a_1) < -2E[log(z_t)]$.
(c) When $E[log(a_1 z_t^2 + b_1)] \geq 0$, Kluppelberg et al. (2004) showed that h_t converges in probability toward $+\infty$: nonstationary GARCH(1,1) processes are explosive.

In the GARCH setting, Bollerslev (1986) proves a simple necessary and sufficient condition for the existence of a second order stationary solution that is non anticipative. Contrary to the preceding proposition, this condition only involves the GARCH structure parameters and is completely independent of the distribution of z_t:

Proposition 2.2.2 *The model* (2.9) *has a unique second order stationary and non anticipative solution if and only if* $a_1 + b_1 < 1$.

Remark 2.2.3

(a) From Remark 2.2.2, the condition $a_1 + b_1 < 1$ implies the strict stationarity of the process, thus, to prove the existence of a second order stationary solution in this case it is sufficient to prove that the unique strictly stationary solution given by Proposition 2.2.1 fulfills $E[X_t^2] = E[h_t] < \infty$. By the Fubini-Tonelli's theorem we deduce from expression (2.13) that

$$E[a_0 + a_0 \sum_{j=1}^{\infty} \prod_{i=1}^{j} (a_1 z_{t-i}^2 + b_1)] = a_0 + a_0 \sum_{j=1}^{\infty} (a_1 + b_1)^j.$$

When $a_1 + b_1 < 1$, one obtains

$$E[X_t^2] = \frac{a_0}{1 - (a_1 + b_1)}. \tag{2.14}$$

(b) When $a_1 + b_1 < 1$, the unique second order stationary and non anticipative solution X_t is a weak white noise because $E[X_{t+h} X_t] = E[X_{|h|} X_0] = 0$.

We have already seen (Remark 2.2.2), that the second order stationary condition implies the strict stationarity one. Nevertheless, the converse is false in general. In fact, when $a_1 + b_1 = 1$, condition (2.12) is fulfilled when $P(z_t^2 = 1) < 1$ (using the strict Jensen inequality). In this case the model is called an integrated GARCH(1,1) (IGARCH(1,1))

$$\begin{cases} X_t = \sqrt{h_t} z_t \\ h_t = a_0 + a_1 X_{t-1}^2 + (1 - a_1) h_{t-1} \end{cases} \tag{2.15}$$

and has the particular property to have finite second order conditional moments and an infinite unconditional one (see Fig. 2.6). It was introduced by Engle and Bollerslev (1986) that first remarked that, in the applications, it has often been observed that the estimated sum of the parameters of a GARCH(1,1) model $a_1 + b_1$ is close to unity. One popular version, implemented in particular in JP Morgan RiskMetrics^{TM}, is the exponentially weighted moving average (EWMA) model for which the restriction $a_0 = 0$ is imposed making the process h_t a martingale. The IGARCH(1,1) models with or without trends are therefore part of a wider class of models characterized by the persistent variancc property that current information remains fundamental for any forecast horizon as we are going to see below.

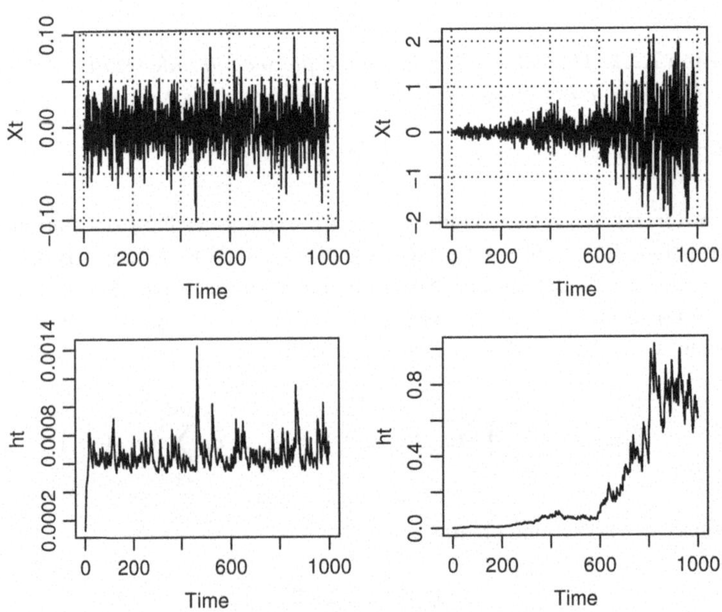

Fig. 2.6 Simulations of size 1,000 of a GARCH(1,1) process X_t and its associated conditional variance h_t with $a_0 = 10^{-5}$, $a_1 = 0.05$, $b_1 = 0.8$, $z_t \hookrightarrow \mathcal{N}(0,1)$ (*left column*) and of an IGARCH(1,1) process X_t and its associated conditional variance h_t with $a_0 = 10^{-5}$, $a_1 = 0.05$, $b_1 = 0.95$ and $z_t \hookrightarrow \mathcal{N}(0,1)$ (*right column*). Both models are strictly stationary but the IGARCH(1,1) process does not admit a second order moment as suggested by the presence of larger absolute values. In each case, large absolute values are not uniformly distributed across time but come in bulk

Under the finite second order moment condition, it is relatively straightforward to obtain optimal (in the sense of conditional expectation) volatility forecasts given the past observations (see for example Baillie and Bollerslev 1992). This property is particularly important when computing Value at Risks under GARCH dynamics hypotheses (see Engle 2001). In fact, we deduce from the linear conditional variance structure

$$h_t = a_0 + a_1 X_{t-1}^2 + b_1 h_{t-1},$$

that the one step predictor of the conditional variance is a weighted combination of the unconditional variance and the last period conditional variance:

$$E[h_{t+1} \mid \mathcal{F}_t] = a_0 + (a_1 + b_1)h_t = \frac{a_0}{1-(a_1+b_1)} + (a_1+b_1)(h_t - \frac{a_0}{1-(a_1+b_1)}).$$

Using the preceding formula recursively, we obtain the k-step ahead forecast

$$E[h_{t+k} \mid \mathcal{F}_t] = \frac{a_0}{1-(a_1+b_1)} + (a_1+b_1)^k (h_t - \frac{a_0}{1-(a_1+b_1)}) \tag{2.16}$$

that is a monotonic function of the forecast horizon. Moreover, in this case where $a_1 + b_1 < 1$, we remark that $E[h_{t+k} \mid \mathcal{F}_t]$ exponentially converges toward the unconditional variance $\frac{a_0}{1-(a_1+b_1)}$ when the forecast horizon goes to infinity and so that the influence of h_t becomes more and more negligible. However, when $a_1 + b_1 = 1$ current information remains important for forecasts of all horizons because

$$E[h_{t+k} \mid \mathcal{F}_t] = ka_0 + h_t.$$

The quantity $a_1 + b_1$, is called the persistence parameter of variance as a measure of the impact of the actual conditional variance on the predictions. This parameter also plays a fundamental role in the autocorrelation structure of the squares of a GARCH process as we are going to see in the next section.

2.2.4 Covariance Structure of the Squares

We have seen previously that, under the finite second order moment condition $a_1 + b_1 < 1$, X_t is a weak white noise compatible with the absence of serial correlations observed in financial time series of log-returns. Now, we are going to show that, under mild and explicit integrability conditions, the autocorrelations of the squares of a GARCH(1,1) process are nonnegative and always decrease (see Fig. 2.3). If $E[X_t^4] < +\infty$, we define $\forall h \in \mathbb{N}^*$,

$$\rho_2(h) = \frac{Cov[X_t^2, X_{t-h}^2]}{Var[X_t^2]}. \tag{2.17}$$

Supposing that $E[z_t^4] < +\infty$ we obtain from

$$E[X_t^4] = E[E[X_t^4 \mid \mathcal{F}_{t-1}]] = E[h_t^2]E[z_t^4] \tag{2.18}$$

that $E[X_t^4] < +\infty$ if and only if $E[h_t^2] < +\infty$. Representation (2.13) gives us

$$E[h_t^2] = E[(a_0 \sum_{j=0}^{\infty} \underbrace{\prod_{i=1}^{j} (a_1 z_{t-i}^2 + b_1)}_{c_j})^2]. \tag{2.19}$$

Using the fact that the z_t are i.i.d random variables with mean zero and variance one

$$E[c_j^2] = (k[z_t]a_1^2 + b_1^2 + 2a_1b_1)^j$$

and

$$E[c_j c_k] = (k[z_t]a_1^2 + b_1^2 + 2a_1b_1)^k (a_1 + b_1)^{j-k} \text{ for } j > k$$

where $k[z_t]$ is the kurtosis of z_t. Consequently, $E[h_t^2] < +\infty$ if and only if

$$k[z_t]a_1^2 + b_1^2 + 2a_1b_1 < 1 \tag{2.20}$$

and in this case

$$E[X_t^4] = a_0^2 k[z_t] \frac{1 + a_1 + b_1}{(1 - a_1 - b_1)(1 - k[z_t]a_1^2 - b_1^2 - 2a_1b_1)} \tag{2.21}$$

and thus

$$k[X_t] = k[z_t] \frac{1 - (a_1 + b_1)^2}{(1 - k[z_t]a_1^2 - b_1^2 - 2a_1b_1)}. \tag{2.22}$$

From Eq. (2.3), the kurtosis of X_t is strictly greater than the z_t one: GARCH models are compatible with the leptokurticity of financial time series mentioned in Sect. 2.1.3 even if the residuals are Gaussian. From the preceding computations we obtain the following proposition:

Proposition 2.2.3 *When* $E[z_t^4] < +\infty$, *the model* (2.9) *has a finite fourth order moment if and only if* $k[z_t]a_1^2 + b_1^2 + 2a_1b_1 < 1$ *and in this case*

$$E[X_t^4] = a_0^2 k[z_t] \frac{1 + a_1 + b_1}{(1 - a_1 - b_1)(1 - k[z_t]a_1^2 - b_1^2 - 2a_1b_1)}.$$

Remark 2.2.4

(a) When z_t follows a $\mathcal{N}(0,1)$ the preceding condition becomes $3a_1^2 + b_1^2 + 2a_1b_1 < 1$ and in this case

$$E[X_t^4] = 3a_0^2 \frac{1 + a_1 + b_1}{(1 - a_1 - b_1)(1 - 3a_1^2 - b_1^2 - 2a_1b_1)}.$$

We have represented in Fig. 2.7 the domains of existence of variance and kurtosis in this case.

(b) The fourth moment structure of a general GARCH(p,q) model has been studied in He and Terasvirta (1999a) and an explicit computation of the kurtosis is provided in Bai et al. (2004). For higher order moments, necessary and sufficient conditions may be found in Ling and McAleer (2002a). When they exist, the odd order moments of a GARCH process are equal to zero when z_t follows a symmetric distribution. In the presence of asymmetry, closed-form formulas are difficult to obtain in practice.

Under condition (2.20), we compute, following He and Terasvirta (1999a),

$$\rho_2(h) = \frac{Cov[X_t^2, X_{t-h}^2]}{Var[X_t^2]}.$$

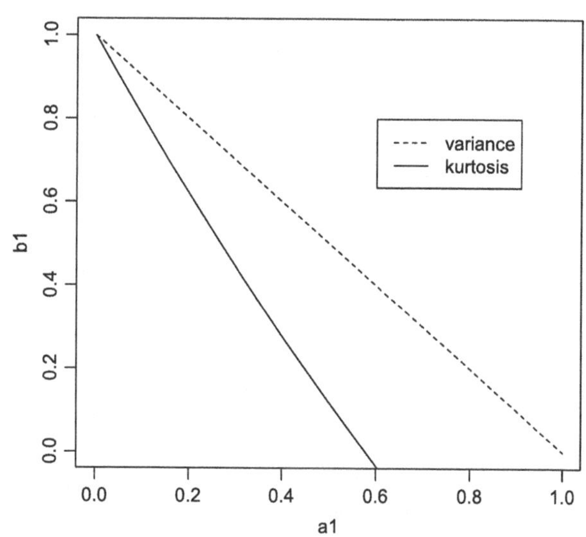

Fig. 2.7 Restrictions on parameters implied by the existence of second and fourth order moments for a GARCH(1,1) model with Gaussian errors. Limit conditions $a_1 + b_1 = 1$ and $3a_1^2 + b_1^2 + 2a_1b_1 = 1$ have been represented

In fact, using Eq. (2.11), we have

$$X_t^2 X_{t-h}^2 = \underbrace{\left(a_0 \sum_{j=0}^{h-1} \prod_{k=1}^{j} A(z_{t-k}) + \prod_{k=1}^{h} A(z_{t-k})\, h_{t-h} \right)}_{h_t} z_t^2 h_{t-h} z_{t-h}^2 \tag{2.23}$$

where $A(x) = a_1 x^2 + b_1$. The conditional variance h_{t-h} being $\sigma(z_u; u \leq t-h-1)$ measurable we have, using the independence of the z_t,

$$E[X_t^2 X_{t-h}^2] = a_0 E[h_t] \frac{1-(a_1+b_1)^h}{1-(a_1+b_1)} + (a_1+b_1)^{h-1} E[h_t^2](a_1 k[z_t] + b_1). \tag{2.24}$$

Thus, the preceding relation, combined with

$$E[h_t] = \frac{a_0}{1-(a_1+b_1)} \text{ and } E[h_t^2] = a_0^2 \frac{1+a_1+b_1}{(1-a_1-b_1)(1-k[z_t]a_1^2 - b_1^2 - 2a_1 b_1)},$$

permits to obtain, from the definition of ρ_2, the following expression after some additional algebraic manipulations:

Proposition 2.2.4 $\forall h \in \mathbb{N}^*$,

$$\rho_2(h) = (a_1+b_1)^{h-1} \frac{(a_1 k[z_t] + b_1)(1-(a_1+b_1)^2) - (a_1+b_1)(1-k[z_t]a_1^2 - b_1^2 - 2a_1 b_1)}{k[z_t](1-(a_1+b_1)^2) - (1-k[z_t]a_1^2 - b_1^2 - 2a_1 b_1)}. \tag{2.25}$$

In particular when z_t follows a $\mathcal{N}(0,1)$,

$$\rho_2(h) = (a_1+b_1)^{h-1} \frac{a_1(1-b_1(a_1+b_1))}{1-(a_1+b_1)^2 + a_1^2}. \tag{2.26}$$

Remark 2.2.5 In the Gaussian case (see Bollerslev 1986), it is easy to derive from the condition $3a_1^2 + b_1^2 + 2a_1 b_1 < 1$ that

$$\frac{a_1(1-b_1(a_1+b_1))}{1-(a_1+b_1)^2 + a_1^2} \geq 0. \tag{2.27}$$

In particular the autocorrelation function of the squares is always nonnegative. This property may be generalized to any GARCH(p,q) model (see Francq and Zakoian 2010, Proposition 2.2) and is compatible with what is generally observed for log-returns of financial assets (see Fig. 2.3). In the GARCH(1,1) case (and only in this case), we deduce from Eq. (2.25) that ρ_2 exponentially decays at the speed of the persistence parameter of variance $a_1 + b_1$ from the initial level $\rho_2(1)$.

2.2.5 Why We Need More: Kurtosis and Asymmetry in a GARCH(1,1) Model

In this section, we underline some drawbacks of the GARCH(1,1) model with Gaussian errors to explain why extended GARCH structures and non-Gaussian distributions will be considered later on. More precisely, the absence of excess kurtosis and the symmetry around zero are the two fundamental properties of the Gaussian distribution that are particularly questioned when combined with the classical GARCH(1,1) process.

In the Gaussian case, when it exists, the autocorrelation function (2.26) may be written as a function of the unconditional kurtosis (2.22):

$$\rho_2(h) = (a_1 + b_1)^{h-1} \left(\frac{b_1(k[X_t] - 3)}{3(k[X_t] - 1)} + a_1 \right). \tag{2.28}$$

In particular, we deduce from

$$\frac{\partial \rho_2(1)}{\partial k[X_t]} = \frac{6b_1}{9(k[X_t] - 1)^2} \geq 0$$

that the initial level of the autocorrelation function is an increasing function of the unconditional kurtosis $k[X_t]$ (it first increases rapidly and then gradually slows down) having a_1 as a minimum (for $k[X_t] = 3$). This relation explains why, in general, this model is not capable of producing realizations with both high kurtosis and low, slowly decaying autocorrelations. To illustrate this phenomena, we have represented in Fig. 2.8 the isoquants of the pair (unconditional kurtosis, $\rho_2(1)$) for various values of $a_1 + b_1$. As observed in Terasvirta (2009) for a large panel of financial assets (and other GARCH parameterizations), empirical values of the pair (unconditional kurtosis, $\rho_2(1)$) are below the lowest isoquant obtained when the persistence parameter $a_1 + b_1 \to 1$ (explaining why GARCH(1,1) models estimated on financial datasets are very closed to IGARCH(1,1) specifications). In the GARCH(1,1) setting, we may overcome this problem using for the z_t a leptokurtic distribution (see Terasvirta and Zhao 2011). Several possible choices will be presented in Sect. 2.4 as an alternative to the Gaussian modeling.

From Table 2.2, the S&P500 log-returns from January 3, 1990 to April 18, 2012 are slightly negatively skewed. Nevertheless, once the distribution of z_t is symmetric around zero we obtain

$$E[X_t^3] = E[h_t^{\frac{3}{2}}]E[z_t^3] = 0,$$

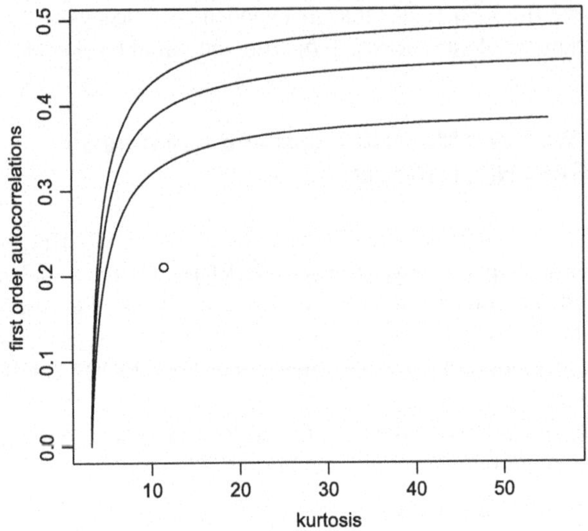

Fig. 2.8 Representation of the pair (unconditional kurtosis, $\rho_2(1)$) for various values of $a_1 + b_1$ for a GARCH(1,1) when z_t follows a $\mathcal{N}(0, 1)$. Isoquants from highest to lowest: $a_1 + b_1 = 0.9, 0.95, 0.99$. When the unconditional kurtosis (resp. $\rho_2(1)$) is fixed, $\rho_2(1)$ (resp. the unconditional kurtosis) is a decreasing (resp. an increasing) function of $a_1 + b_1$. The *circle* represents the pair (unconditional kurtosis, $\rho_2(1)$) for the S&P500 log-returns dataset considered in Sect. 2.1. It is below the lowest isoquant $a_1 + b_1 = 0.99$. This phenomena is also present for other, even asymmetric, GARCH type models (see Malmsten and Terasvirta 2004)

thus unconditional asymmetry is excluded. Moreover, regarding the leverage effect observed in Sect. 2.1.2, it may be described by the quantity

$$Cov[X_t - X_{t-1}, h_{t+1} - h_t \mid \mathcal{F}_{t-1}]$$

measuring the impact (empirically observed as negative) of the present variations of the log-returns X_t on the future variations of the conditional variance h_t. In the GARCH(1,1) setting,

$$\begin{aligned} Cov[X_t - X_{t-1}, h_{t+1} - h_t \mid \mathcal{F}_{t-1}] &= Cov[X_t, h_{t+1} \mid \mathcal{F}_{t-1}] \\ &= Cov[X_t, a_0 + a_1 X_t^2 + b_1 h_t \mid \mathcal{F}_{t-1}] \\ &= a_1 Cov[X_t, X_t^2 \mid \mathcal{F}_{t-1}] \\ &= a_1 h_t^{\frac{3}{2}} E[z_t^3]. \end{aligned}$$

Once again, the leverage is equal to zero if z_t is symmetric and compatible with empirical observations if $E[z_t^3] < 0$. However, starting from a skewed error is not the only way to deal with the leverage effect stylized fact. In fact, since

$$Cov[X_t - X_{t-1}, h_{t+1} - h_t \mid \mathcal{F}_{t-1}] = 0 \Leftrightarrow Cov[X_t^+, h_{t+1} \mid \mathcal{F}_{t-1}] = Cov[X_t^-, h_{t+1} \mid \mathcal{F}_{t-1}]$$

where

$$X_t^+ = Max(X_t, 0) \text{ and } X_t^- = Max(-X_t, 0)$$

it is possible to modify the conditional volatility structure of the model to cope with asymmetric reactions of h_t to negative or positive values of X_t (that is to bad or good financial news). This is the aim of the next section.

2.3 Asymmetric Extensions

The original GARCH(1,1) specification of the conditional variance (2.9), has been generalized and extended in various directions to increase the practical flexibility of the model and incorporate, in particular, asymmetry properties [see e.g. Terasvirta (2009) for a recent survey and Bollerslev (2011) for an encyclopedic type reference guide to the long list of ARCH acronyms that have been used in the literature]. Indeed, the classical GARCH parameterization $h_t = a_0 + a_1 X_{t-1}^2 + b_1 h_{t-1}$ assumes by construction that the current conditional variance is a linear function of the squared past innovation X_{t-1}^2. Even if it allows the volatility clustering effect to be reproduced in a very simple way, this parametric form constrains the response of the variance to a shock to be independent of the sign of this shock. To take into account that the volatility increase due to a price decrease (negative log-returns) is generally stronger than the one resulting from a price increase (positive log-returns) of the same magnitude, both Engle (1990) and Sentana (1995) introduced the following asymmetric GARCH model:

$$h_t = a_0 + a_1 (X_{t-1} + \gamma)^2 + b_1 h_{t-1} \tag{2.29}$$

where γ is a strictly negative constant. In this case, the center of symmetry of the response to shocks is shifted away from zero to capture the leverage effect: the negative coefficient γ entails that a negative X_{t-1} has a bigger impact on h_t than a positive one. We have plotted in Fig. 2.9 the so-called news impact curves introduced by Engle and Ng (1993) representing the impact of X_{t-1} on h_t for the GARCH(1,1) and its asymmetric counterpart (2.29). These curves measure how new information is incorporated in volatility estimates. For the asymmetric GARCH model this curve is asymmetric and centered at $X_{t-1} = -\gamma$ which is to the right of the origin when imposing $\gamma < 0$ in accordance with the empirical properties of financial time

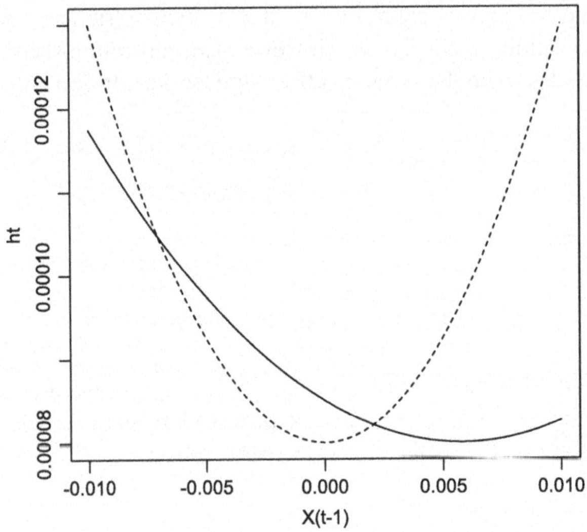

Fig. 2.9 News impact curves of the GARCH(1,1) model (*dashed line*) and the asymmetric GARCH model (*solid line*). The equation for the GARCH(1,1) news impact curve is $h_t = a_0 + a_1 X_{t-1}^2 + b_1\sigma^2$ with $a_0 = 10^{-5}, a_1 = 0.2$ and $b_1 = 0.7$ and where $\sigma^2 = \frac{a_0}{1-(a_1+b_1)} = 10^{-4}$ is the unconditional variance. The equation for the asymmetric GARCH news impact curve is $h_t = \tilde{a_0} + \tilde{a_1}(X_{t-1} + \gamma)^2 + \tilde{b_1}\tilde{\sigma}^2$ with $\tilde{a_0} = 10^{-5}, \tilde{a_1} = 0.15, \tilde{b_1} = 0.7$ and $\gamma = -5.8 \times 10^{-3}$ and where $\tilde{\sigma}^2 = \frac{\tilde{a_0}+\tilde{a_1}\gamma^2}{1-(\tilde{a_1}+\tilde{b_1})} = 10^{-4}$ is the unconditional variance. For both models, the coefficients have been chosen to fulfill the second order stationarity conditions $a_1 + b_1 < 1$ and $\tilde{a_1} + \tilde{b_1} < 1$ and to make the unconditional variances σ^2 and $\tilde{\sigma}^2$ equal

series of log-returns. Nevertheless, we can also remark, from this figure, one of the major drawback of this approach: small prices increases may produce a smaller volatility than log-returns equal to zero (the minimum of the news impact curve being achieved for $X_{t-1} = -\gamma$).

To avoid this counterintuitive property, we introduce, in the following, three particular GARCH extensions accommodating with asymmetric volatility responses, namely, the GJR GARCH model introduced by Glosten et al. (1993), the exponential GARCH (EGARCH) model of Nelson (1991) and the asymmetric power ARCH (APARCH) model developed by Ding et al. (1993). These models probably represent the most popular GARCH alternatives used in practice and, contrary to the asymmetric GARCH, are not built on shifted news impact curves but on curves centered at 0 allowing slopes of different magnitude on either side of the origin. We refer the reader to Hentschel (1995) for a unified framework, based on Box–Cox transforms of the conditional standard deviation, containing these three specifications as important special cases.

2.3.1 GJR Model

To take into account the dependence of a coefficient of the volatility structure on one particular event (here $X_{t-1} < 0$) it is natural to introduce in the GARCH specification the indicator function of this event. This idea was used by Glosten et al. (1993) to modify the classical GARCH(p,q) model (2.8) in the following way:

Definition 2.3.1 The process $(X_t)_{t\in\mathbb{Z}}$ is a GJR-GARCH process of order $(p,q) \in (\mathbb{N}^*)^2$ if

$$\begin{cases} X_t = \sqrt{h_t} z_t \\ h_t = a_0 + \sum_{i=1}^{p} (a_i X_{t-i}^2 + \gamma_i \mathbf{1}_{X_{t-i}<0} X_{t-i}^2) + \sum_{j=1}^{q} b_j h_{t-j} \end{cases} \tag{2.30}$$

where the $(z_t)_{t\in\mathbb{Z}}$ are i.i.d random variables with $E[z_t] = 0$ and $Var[z_t] = 1$ and where $\mathbf{1}_{X_{t-i}<0}$ is the indicator function such that $\mathbf{1}_{X_{t-i}<0} = 1$ if $X_{t-i} < 0$ and $\mathbf{1}_{X_{t-i}<0} = 0$ otherwise. The a_i, γ_i and b_j are nonnegative constants such that $a_0 > 0$.

When the γ_i are equal to zero we recover the symmetric GARCH(p,q) model. In practice, we will consider the simplest model of order $(1,1)$ (simply called the GJR model in the following) such that

$$\begin{aligned} h_t &= a_0 + a_1 X_{t-1}^2 + \gamma \mathbf{1}_{X_{t-1}<0} X_{t-1}^2 + b_1 h_{t-1} \\ &= a_0 + a_1 X_{t-1}^2 + \gamma Max(0, -X_{t-1})^2 + b_1 h_{t-1} \end{aligned}$$

or equivalently

$$h_t = a_0 + a_+ X_{t-1}^2 \mathbf{1}_{X_{t-1}\geq 0} + a_- \mathbf{1}_{X_{t-1}<0} X_{t-1}^2 + b_1 h_{t-1} \tag{2.31}$$

where $a_+ = a_1$ and $a_- = a_1 + \gamma$. Figure 2.10 depicts the difference between the GJR and the asymmetric GARCH models in terms of news impact curves proving that the asymmetry present in both models comes from radically different sources: in the GJR model, the autoregressive coefficient depends on the sign of X_{t-1} while the asymmetric GARCH model shifts the news impact curve of the GARCH(1,1) process away from zero.

From the relation

$$h_t = a_0 + B(z_{t-1}) h_{t-1}$$

where

$$B(x) = (a_1 + \gamma \mathbf{1}_{x<0}) x^2 + b_1,$$

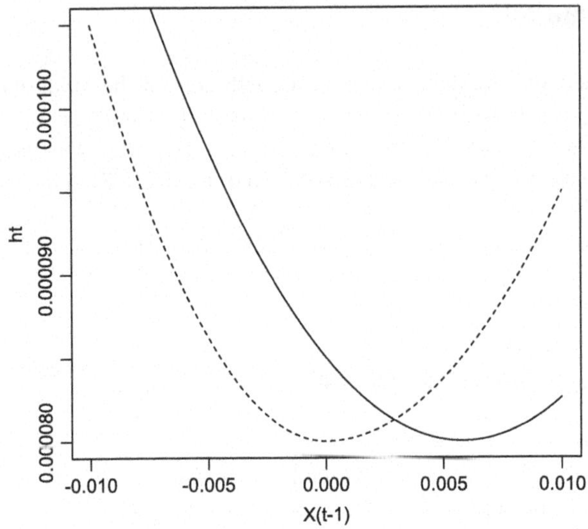

Fig. 2.10 News impact curves of the GJR model (*dashed line*) and the asymmetric GARCH model (*solid line*). The equation for the GJR news impact curve is $h_t = a_0 + a_1 X_{t-1}^2 + \gamma Max(0, -X_{t-1})^2 + b_1\sigma^2$ with $a_0 = 10^{-5}$, $a_1 = 0.15$, $\gamma = 0.1$ and $b_1 = 0.7$ and where $\sigma^2 = \frac{a_0}{1-(a_1+b_1+0.5\gamma)} = 10^{-4}$ is the unconditional variance. The equation for the asymmetric GARCH news impact curve is $h_t = \tilde{a_0} + \tilde{a_1}(X_{t-1} + \tilde{\gamma})^2 + \tilde{b_1}\tilde{\sigma}^2$ with $\tilde{a_0} = 10^{-5}$, $\tilde{a_1} = 0.15$, $\tilde{b_1} = 0.7$ and $\tilde{\gamma} = -5.8 \times 10^{-3}$ and where $\tilde{\sigma}^2 = \frac{\tilde{a_0}+\tilde{a_1}\tilde{\gamma}^2}{1-(\tilde{a_1}+\tilde{b_1})} = 10^{-4}$ is the unconditional variance. For both models, the coefficients have been chosen to fulfill the second order stationarity conditions $a_1 + b_1 + 0.5\gamma < 1$ and $\tilde{a_1} + \tilde{b_1} < 1$ and to make the unconditional variances σ^2 and $\tilde{\sigma}^2$ equal

that is the analogous of Eq. (2.10), it is possible to study the statistical properties of the GJR model along the lines of Sects. 2.2.3 and 2.2.4. In fact, in Francq and Zakoian (2006), a necessary and sufficient condition for the existence and the uniqueness of a strictly stationary and non anticipative solution is given by

$$E[log(B(z_{t-1}))] < 0$$

that is noneother than condition (2.12) when $\gamma = 0$. Conditions of existence of finite moments and covariance structure of the squares may be found in He and Terasvirta (1999b) and Ling and McAleer (2002b). When the distribution of z_t is symmetric (as in the Gaussian case), a second order stationary solution exists if and only if $a_1 + b_1 + 0.5\gamma < 1$ and in this case the unconditional variance of X_t is given by $\frac{a_0}{1-(a_1+b_1+0.5\gamma)}$. In particular, the admissible region for the pairs (a_1, b_1) allowing for second order stationarity of the GJR model is smaller than for the GARCH(1,1) one. These restrictions on the parameters are amplified if we impose the existence of higher order moments. For example, in the Gaussian case, the finite unconditional kurtosis condition is given by

$$b_1^2 + 2a_1b_1 + 3a_1^2 + b_1\gamma + 3a_1\gamma + \frac{3\gamma^2}{2} < 1,$$

thus, the GJR dynamics may be heavily limited in this case (see Rodriguez and Ruiz 2012). In the next section, a different approach of asymmetry for GARCH type structures is presented to be more flexible to represent leverage effect and simultaneously satisfy the restrictions for positivity and existence of the kurtosis.

Remark 2.3.1

(a) The Threshold GARCH model introduced by Zakoian (1994) is a slight different version of the GJR model where the volatility dynamics is specified in terms of conditional standard deviation instead of conditional variance:

$$\begin{cases} X_t = \sigma_t z_t \\ \sigma_t = a_0 + a_+ X_{t-1} \mathbf{1}_{X_{t-1} \geq 0} + a_- \mathbf{1}_{X_{t-1} < 0} X_{t-1} + b_1 \sigma_{t-1}. \end{cases} \tag{2.32}$$

In this model, imposing the positivity of σ_t is not required since the conditional variance is σ_t^2 and the parameters (a_0, a_-, a_+, b_1) are a priori unconstrained. Nevertheless, the conditions for the existence of a second order stationarity solution (that unfortunately depend on the distribution of z_t) and of a finite kurtosis are more complex than in the GJR case.

(b) Several extensions of the first order GJR model have been proposed in the literature to allow for more complex asymmetric responses. For example, in Fornari and Mele (1996) the moving average coefficient b_1 in (2.31) depends on the event $X_{t-1} < 0$ while a smooth transition between the states $X_{t-1} < 0$ and $X_{t-1} \geq 0$ (instead of the simple indicator function) is proposed in Gonzalez-Rivera (1998).

2.3.2 EGARCH Model

To avoid positivity constraints on the GARCH coefficients and allow for leverage effect, Nelson (1991) introduced the so-called exponential GARCH model presented in the next definition in its general form:

Definition 2.3.2 The process $(X_t)_{t \in \mathbb{Z}}$ is an exponential GARCH (EGARCH) process of order $(p, q) \in (\mathbb{N}^*)^2$ if

$$\begin{cases} X_t = \sqrt{h_t} z_t \\ log(h_t) = w + \sum_{i=1}^{p} \alpha_i g(z_{t-i}) + \sum_{j=1}^{q} \beta_j log(h_{t-j}) \end{cases} \tag{2.33}$$

where

$$g(z_{t-i}) = \theta z_{t-i} + \xi(|z_{t-i}| - E[|z_{t-i}|])$$

and where the $(z_t)_{t \in \mathbb{Z}}$ are i.i.d random variables with $E[z_t] = 0$ and $Var[z_t] = 1$.

When $p = q = 1$, we have

$$log(h_t) = w + \alpha_1(\theta z_{t-1} + \xi(|z_{t-1}| - E[|z_{t-1}|])) + \beta_1 log(h_{t-1})$$

and defining $a_0 = w - \alpha_1 \xi E[|z_{t-1}|]$, $a_1 = \alpha_1 \xi$, $\gamma = -\frac{\theta}{\xi}$ and $b_1 = \beta_1$ we obtain the following simplified form that will be used in the empirical Chap. 4:

$$log(h_t) = a_0 + a_1(|z_{t-1}| - \gamma z_{t-1}) + b_1 log(h_{t-1}). \tag{2.34}$$

Contrary to the preceding GARCH specifications, the volatility dynamics (2.34) has a multiplicative form instead of a linear one

$$h_t = e^{a_0 + a_1(|z_{t-1}| - \gamma z_{t-1})} h_{t-1}^{b_1}$$

and there are no restrictions on the parameters to guarantee the positivity of the conditional variance due to the presence of the exponential function. Moreover, when $z_{t-1} > 0$ (resp. $z_{t-1} < 0$), $log(h_t)$ is a linear function of z_{t-1} with slope $a_1(1-\gamma)$ (resp. $-a_1(1+\gamma)$). Thus, if $\gamma \neq 0$ the volatility responds asymmetrically to rises and falls in stock prices and when $|\gamma| < 1$ this asymmetry is compatible with empirical leverage effects.

Remark 2.3.2 In Eq. (2.34) the conditional variance is a function of the past standardized innovation z_{t-1} instead of a function of the past innovation X_{t-1} as in the GARCH(1,1) or the GJR models. It may intuitively explain why the EGARCH model, contrary to the GARCH(1,1) one, usually presents a lack of volatility clustering as theoretically proved in Mikosch and Rezapur (2012).

From the autoregressive volatility structure (2.34) and the fact that the random variables $(|z_{t-1}| - \gamma z_{t-1})$ are i.i.d, it is relatively simple to prove stationarity and finite moments properties of the EGARCH model. In fact, we deduce by a recursive argument that $\forall h \in \mathbb{N}^*$,

$$log(h_t) = a_0 \sum_{k=0}^{h-1} b_1^k + a_1 \sum_{k=1}^{h} b_1^{k-1}(|z_{t-k}| - \gamma z_{t-k}) + b_1^h log(h_{t-h}) \tag{2.35}$$

and when $|b_1| < 1$ there exists a unique strictly stationary solution given by

$$log(h_t) = \frac{a_0}{1-b_1} + a_1 \sum_{k=1}^{+\infty} b_1^{k-1}(|z_{t-k}| - \gamma z_{t-k}). \tag{2.36}$$

Using that, for a positive integer p,

$$E[X_t^{2p}] = E[z_t^{2p}]E[h_t^p],$$

necessary and sufficient conditions for the existence of a moment of order $2p$ are obtained from the finiteness of the quantity $E[h_t^p]$ that may be expressed from Eq. (2.36) and independence by

$$E[h_t^p] = e^{\frac{pa_0}{1-b_1}} \prod_{k=1}^{+\infty} E[e^{pa_1 b_1^{k-1}(|z_{t-k}|-\gamma z_{t-k})}].$$

This approach was proposed in He et al. (2002) where explicit forms (involving infinite products) of the variance, the kurtosis and the autocorrelation function of the squares are given. In particular, when z_t follows a $\mathcal{N}(0, 1)$, moments of all order exist under the simple global condition $|b_1| < 1$. Contrary to what happens for the GARCH models previously mentioned, the moment conditions do not become more and more stringent for higher and higher even moments. This nice analytic property also proves that the empirical leptokurticity property of X_t may be more difficult to capture with this parameterization.

2.3.3 APARCH Model

In Ding et al. (1993), the authors remark that, for a large number of financial time series, the autocorrelation functions of the power transformed log-returns $|X_t|^\delta$ peak around $\delta = 1$ (see Table 2.3). They called this property the Taylor effect because it was first observed in Taylor (1986) that autocorrelations are in general greater for absolute returns $|X_t|$ than for squared ones $|X_t|^2$. This empirical feature is now commonly added to the list of stylized facts that characterize stock return dynamics (see e.g. Cont 2001; Granger 2005).

From this observation, Ding et al. (1993) advocate that there are no strong empirical reasons to only consider conditional variances dynamics ($\delta = 2$) instead of arbitrary powered ones. In fact, the common use of conditional second order

Table 2.3 Autocorrelations of $|X_t|^\delta$ for the log-returns of the S&P500 index from January 3, 1990 to April 18, 2012

	Lag $h = 1$	$h = 2$	$h = 3$	$h = 4$	$h = 5$	$h = 10$	$h = 15$	$h = 20$	$h = 30$
$\delta = 0.1$	0.104	0.094	0.101	0.091	0.114	0.104	0.103	0.082	0.093
$\delta = 0.5$	0.191	0.202	0.201	0.200	0.228	0.195	0.184	0.170	0.158
$\delta = 0.8$	0.227	0.250	0.238	0.234	0.268	0.223	0.207	0.196	0.171
$\delta = 0.9$	0.237	0.263	0.246	0.241	0.279	0.228	0.210	0.200	0.172
$\delta = 1$	0.245	0.275	0.251	0.245	0.287	0.229	0.211	0.201	0.170
$\delta = 1.1$	0.251	0.285	0.253	0.245	0.293	0.227	0.208	0.198	0.166
$\delta = 1.2$	0.253	0.292	0.251	0.241	0.295	0.221	0.202	0.192	0.158
$\delta = 1.5$	0.236	0.289	0.218	0.203	0.279	0.179	0.164	0.154	0.122
$\delta = 2$	0.144	0.210	0.113	0.099	0.189	0.083	0.075	0.067	0.051
$\delta = 3$	0.029	0.069	0.015	0.011	0.060	0.010	0.007	0.005	0.004

moments implicitly comes from Gaussian assumptions under which distributions are perfectly characterized by the two first moments. In this way, a new class of asymmetric GARCH models was proposed in Ding et al. (1993) involving an additional heteroskedasticity parameter which increases the flexibility of the model:

Definition 2.3.3 The process $(X_t)_{t\in\mathbb{Z}}$ is an asymmetric power ARCH (APARCH) process of order $(p, q) \in (\mathbb{N}^*)^2$ if

$$\begin{cases} X_t = \sqrt{h_t} z_t \\ h_t^{\frac{\Delta}{2}} = a_0 + \sum_{i=1}^{p} a_i(|X_{t-i}| - \gamma_i X_{t-i})^{\Delta} + \sum_{j=1}^{q} b_j h_{t-j}^{\frac{\Delta}{2}} \end{cases} \tag{2.37}$$

where a_i and b_j are nonnegative constants, $a_0 > 0$, $\Delta > 0$, $|\gamma_i| \leq 1$ and where the $(z_t)_{t\in\mathbb{Z}}$ are i.i.d random variables with $E[z_t] = 0$ and $Var[z_t] = 1$.

Remark 2.3.3 The condition $|\gamma_i| \leq 1$ implies that $(|X_{t-i}| - \gamma_i X_{t-i}) > 0$ and since

$$a_i(|X_{t-i}| - \gamma_i X_{t-i})^{\Delta} = a_i|\gamma_i|^{\Delta}(\frac{1}{|\gamma_i|}|X_{t-i}| - sign(\gamma_i)X_{t-i})^{\Delta}$$

it may also be seen as a nonrestrictive identifiability constraint.

When $p = q = 1$, we have

$$h_t^{\frac{\Delta}{2}} = a_0 + a_1(|X_{t-1}| - \gamma X_{t-1})^{\Delta} + b_1 h_{t-1}^{\frac{\Delta}{2}} \tag{2.38}$$

and, as remarked in Ding et al. (1993), this model nests most of the GARCH models with leverage effect described before (see Table 2.4).

Once again, using the linear relation (2.38) fulfilled by $h_t^{\frac{\Delta}{2}}$ we can write

$$h_t^{\frac{\Delta}{2}} = a_0 + C(z_{t-1})h_{t-1}^{\frac{\Delta}{2}}$$

where $C(x) = a_1(|x| - \gamma x)^{\Delta} + b_1$ and, as in the GARCH(1,1) case, a necessary and sufficient condition for the existence and the uniqueness of a non anticipative and strictly stationary solution is given by

$$E[log(C(z_t))] < 0.$$

Table 2.4 Special cases of the APARCH(1,1) model

$\Delta = 2, \gamma = 0$	GARCH model (2.9)
$\Delta = 2, \gamma \neq 0$	GJR model (2.31)
$\Delta = 1, \gamma \neq 0$	TGARCH model (2.32)
$\Delta \to 0$	Log-GARCH model of Pantula (1986)

Contrary to what happens for GARCH and GJR models, this condition depends, even in the symmetric case, on the distribution of z_t. Assuming that $E[|z_t|^\Delta] < \infty$, the strictly stationary solution fulfills $E[X_t^\Delta] < \infty$ if $E[C(z_t)] < 1$ which reduces for symmetric distributions to

$$\frac{E[|z_t|^\Delta]}{2} a_1((1+\gamma)^\Delta + (1-\gamma)^\Delta) + b_1 < 1.$$

The term $E[|z_t|^\Delta]$ may be computed explicitly (in terms of the Euler gamma function) in the Gaussian case (see Ding et al. 1993) and for other classical distributions (see Karanasos and Kim 2006). When $\Delta \geq 2$, the preceding condition is sufficient to ensure the second order stationarity of the solution but is not necessary in general. We can find in He and Terasvirta (1999c) a more complete study of the moment properties of the first order APARCH model but closed-form expressions of unconditional variance and kurtosis or autocorrelation function of the squares are only available for very special cases. Nevertheless, even if the free heteroskedasticity parameter Δ increases the analytic complexity of the model, it allows it to cope with the Taylor effect contrary to what happen for the classical GARCH(1,1) process (see He and Terasvirta 1999b).

2.3.4 Concluding Remarks

Among the large number of alternative GARCH models proposed in the financial econometrics literature, we have presented in this section three very popular parameterizations. They are probably the most widely used in empirical applications to represent the dynamic evolution of volatility with leverage effects. Their analytic properties are, in general, explicitly described at the very least in the case of innovations following a Gaussian distribution. However, in spite of its remarkable mathematical tractability, the Gaussian hypothesis is hardly compatible with the empirical features of financial returns even associated with asymmetric GARCH structures. We have represented in Table 2.5 the descriptive statistics of the standardized residuals obtained from the S&P500 log-returns when the time varying variance has been filtered out using the previous GARCH specifications.

Table 2.5 Descriptive statistics of the standardized residuals of asymmetric GARCH specifications with Gaussian innovations fitted on the log-returns of the S&P500 index from January 3, 1990 to April 18, 2012

Model	Mean	Variance	Skewness	Kurtosis	KS p-value	AD p-value
GJR	−4E−03	1	−0.49	7.74	0	0
EGARCH	−8E−04	0.99	−0.55	8.32	0	0
APARCH	−8E−03	0.99	−0.49	7.62	0	0

The two last columns present the p-values associated to the Kolmogorov–Smirnov and Andersen–Darling tests, comparing the standardized residuals to the Gaussian distribution

Clearly, the skewness and kurtosis levels in the standardized residuals are hardly compatible with the Gaussian hypothesis that is rejected at any level by the Kolmogorov–Smirnov and Andersen–Darling tests whatever the choice of the underlying volatility structure. Even if non negligible proportions of the total asymmetry and leptokurticity have been captured by the leveraged GARCH models (see Table 2.2), the situation can be improved by replacing the normal error distribution by a more fat-tailed and negatively skewed one. In the next section, we present several interesting candidates.

2.4 Conditional Distribution of Returns

GARCH type models coupled with the auxiliary assumption of conditionally Gaussian innovations have empirically shown their inability to take into account all the mass in the tails and the asymmetry that characterize the distribution of daily log-returns. A natural way of accommodating such stylized facts is to specify structures driven by alternative i.i.d. innovations. Historically, among the myriad of possible choices, several interesting distributions were proposed to better account for the deviation of normality: the Student distribution and the generalized error distribution, respectively used by Bollerslev (1987) and Nelson (1991), are symmetric but leptokurtic while the skewed Student distribution introduced by Hansen (1994) also allows for asymmetry. In the following, we present two families of distributions known to fit the statistical behavior of asset returns remarkably: the Generalized Hyperbolic (GH) distributions and the mixture of two Gaussian distributions (MN).

2.4.1 Generalized Hyperbolic Distributions

The Generalized Hyperbolic (GH) distribution introduced in a seminal paper by Barndorff-Nielsen (1977) is a normal variance-mean mixture where the mixing distribution is the generalized inverse Gaussian distribution (see the Monte Carlo methods part of the Mathematical Appendix). Equivalently, for $(\lambda, \alpha, \beta, \delta, \mu) \in \mathbb{R}^5$ with $\delta > 0$ and $\alpha >\mid \beta \mid> 0$, the one dimensional $GH(\lambda, \alpha, \beta, \delta, \mu)$ distribution is defined by the following density function: $\forall x \in \mathbb{R}$,

$$d_{GH}(x, \lambda, \alpha, \beta, \delta, \mu) = \frac{(\sqrt{\alpha^2 - \beta^2}/\delta)^\lambda}{\sqrt{2\pi} K_\lambda(\delta\sqrt{\alpha^2 - \beta^2})} e^{\beta(x-\mu)} \frac{K_{\lambda-1/2}\left(\alpha\sqrt{\delta^2 + (x-\mu)^2}\right)}{\left(\sqrt{\delta^2 + (x-\mu)^2}/\alpha\right)^{1/2-\lambda}}, \tag{2.39}$$

where K_λ is the modified Bessel function of the third kind

$$K_\lambda(x) = \frac{1}{2}\int_0^{+\infty} y^{\lambda-1} e^{-\frac{x}{2}(y+\frac{1}{y})} dy, \; x > 0. \tag{2.40}$$

For $\lambda \in \frac{1}{2}\mathbb{Z}$, the basic properties of the Bessel function (see Abramowitz and Stegun 1964) allow simpler forms for the density to be found. In particular, for $\lambda = 1$, we get the Hyperbolic distribution (HYP) whose log-density is a hyperbola

$$d_{HYP}(x, \alpha, \beta, \delta, \mu) = \frac{\sqrt{\alpha^2 - \beta^2}}{2\delta\alpha K_1(\delta\sqrt{\alpha^2 - \beta^2})} e^{\beta(x-\mu) - \alpha\sqrt{\delta^2 + (x-\mu)^2}}$$

and for $\lambda = -\frac{1}{2}$, we obtain the Normal Inverse Gaussian distribution (NIG) with density

$$d_{NIG}(x, \alpha, \beta, \delta, \mu) = \frac{\alpha\delta}{\pi} \frac{K_1(\alpha\sqrt{\delta^2 + (x-\mu)^2})}{\sqrt{\delta^2 + (x-\mu)^2}} e^{\beta(x-\mu) + \delta\sqrt{\alpha^2 - \beta^2}}$$

which is closed under convolution. This family is also stable under affine transformations, as proved in Barndorff-Nielsen and Blaesild (1981). This property is interesting because in the GARCH setting we will be able to deduce the conditional distribution of the log-returns from the innovations. In particular, when $\alpha^* = \alpha\delta$ and $\beta^* = \beta\delta$, if $X \hookrightarrow GH(\lambda, \alpha^*, \beta^*, \delta, \mu)$, then,

$$\frac{X - \mu}{\delta} \hookrightarrow GH(\lambda, \alpha, \beta, 1, 0).$$

The parameters μ and δ respectively describe the location and the scale. The other parameters are interpreted as follows: β describes the skewness (when $\beta = 0$ the distribution is symmetric) while α drives the kurtosis (see Fig. 2.11).

As already remarked, the parameter λ characterizes interesting subclasses and also influences the tail behavior because up to a multiplicative constant

$$d_{GH}(x, \lambda, \alpha, \beta, \delta, \mu) \approx |x|^{\lambda-1} e^{(\mp\alpha+\beta)x}$$

when $x \to \pm\infty$ (see Barndorff-Nielsen and Blaesild 1981). From Eberlein and Hammerstein (2003), one particularly interesting feature of the GH distribution is that it contains, as limiting cases, distributions that are widely used in financial applications (see Table 2.6).

In spite of the apparent complexity of its density function (2.39), the mean-variance mixture structure of the GH distribution allows for a simple computation of the moment generating function (see Eberlein and Hammerstein 2003) that is given by

$$\mathbb{G}_{GH}(u) = e^{\mu u} \left(\frac{\alpha^2 - \beta^2}{\alpha^2 - (\beta + u)^2} \right)^{\frac{\lambda}{2}} \frac{K_\lambda(\delta\sqrt{\alpha^2 - (\beta + u)^2})}{K_\lambda(\delta\sqrt{\alpha^2 - \beta^2})}, \quad |\beta + u| < \alpha. \tag{2.41}$$

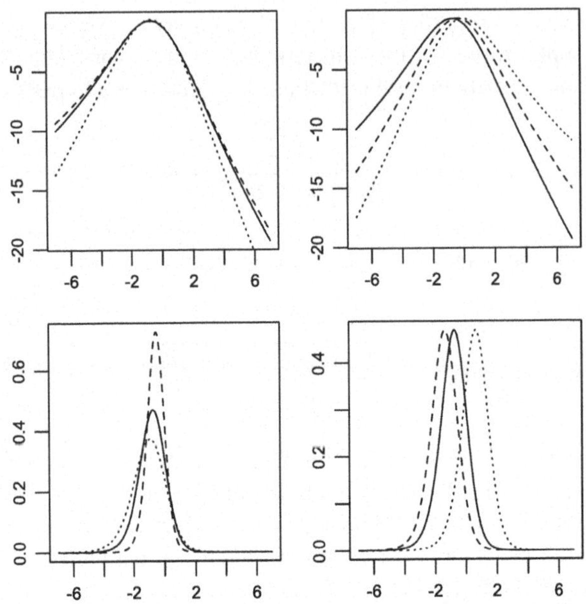

Fig. 2.11 This figure presents the effects of changes in parameters for the log-density (*top*) or the density (*bottom*) of the GH distribution. Starting from the initial values ($\alpha = 0.856, \beta = -0.561, \delta = 2.786, \mu = -0.379, \lambda = -5.614$) (see the EGARCH model parameters in Table 4.7) corresponding to a skewness of −0.36 and a kurtosis of 4.06, one of the first four parameters is changed the others being constant. The *top left figure* presents the effect of changes in the kurtosis parameter α for the log-density ($\alpha = 0.856$ *solid line*, $\alpha = 0.562$ *dashed line*, $\alpha = 2$ *dotted line*). The *top right figure* presents the effect of changes in the skewness parameter β for the log-density ($\beta = -0.561$ *solid line*, $\beta = 0$ *dashed line*, $\beta = 0.561$ *dotted line*). The *bottom left figure* presents the effect of changes in the scale parameter δ for the density ($\delta = 2.786$ *solid line*, $\delta = 1.8$ *dashed line*, $\delta = 3.5$ *dotted line*). The *bottom right figure* presents the effect of changes in the location parameter μ for the density ($\mu = -0.379$ *solid line*, $\mu = -1$ *dashed line*, $\mu = 1$ *dotted line*)

Table 2.6 Non trivial limits of GH distributions

Parameters restrictions	Special case
$\lambda > 0, \delta \to 0$	Variance Gamma (Madan et al. 1998)
$\lambda = 1, \delta \to 0$	Skewed and shifted Laplace (Blaesild 1999)
$\lambda = -\frac{1}{2}, \alpha = \mid \beta \mid \to 0$	Scaled and shifted Cauchy (Blaesild 1999)
$\lambda = -\frac{\delta^2}{2} < 0, \alpha = \mid \beta \mid \to 0$	Shifted Student-t (Blaesild 1999)
$\alpha \to +\infty, \delta \to 0, \alpha\delta^2 \to$ Cste	Shifted Generalized inverse Gaussian (Blaesild 1999)
$\alpha, \delta \to +\infty, \frac{\delta}{\alpha} \to$ Cste	Gaussian (Barndorff-Nielsen 1978)

As a consequence, moments of all orders are finite and, differentiating the preceding expression, we can easily obtain the mean and the variance of a random variable X following a $GH(\lambda, \alpha, \beta, \delta, \mu)$ distribution:

$$E[X] = \mu + \frac{\delta\beta}{\gamma}\frac{K_{\lambda+1}(\delta\gamma)}{K_\lambda(\delta\gamma)}$$
$$Var[X] = \frac{\delta}{\gamma}\frac{K_{\lambda+1}(\delta\gamma)}{K_\lambda(\delta\gamma)} + \frac{\beta^2\delta^2}{\gamma^2}\left(\frac{K_{\lambda+2}(\delta\gamma)K_\lambda(\delta\gamma)-K^2_{\lambda+1}(\delta\gamma)}{K^2_\lambda(\delta\gamma)}\right) \tag{2.42}$$

where $\gamma = \sqrt{\alpha^2 - \beta^2}$ [exact values for the associated skewness and excess kurtosis may also be found in Barndorff-Nielsen and Blaesild (1981)].

First introduced by Barndorff-Nielsen (1977) in connection with the modeling of dune movements, the GH distributions and their subclasses have been suggested, in the last 20 years as a model for financial price processes. In fact, its exponentially decreasing tails seem to fit asset returns dynamics remarkably and are particularly relevant for risk measurement (see among others Barndorff-Nielsen 1995; Eberlein and Keller 1995; Jensen and Lunde 2001; Eberlein and Prause 2002 or Yang 2011). Moreover, the flexible analytic properties of the GH family [infinite divisibility shown in Barndorff-Nielsen and Halgreen (1977), explicit form for the moment generating function (2.41)] make it a key stone of modern continuous time finance for the construction of exponential Lévy models and the associated option pricing approach based on Esscher transform (see e.g. Cont and Tankov 2003). Following Badescu et al. (2011) and Chorro et al. (2012), we will see in Chap. 3 that this distribution is also well adapted for the pricing of financial derivatives in the discrete time GARCH setting.

2.4.2 Mixture of Two Gaussian Distributions

The mixture of two Gaussian distributions (MN) is an interesting flexible alternative to the Gaussian distribution matching financial stylized facts and also a natural competitor of the GH family because it depends on the same number of parameters. For $(\phi, \mu_1, \mu_2, \sigma_1, \sigma_2) \in [0,1] \times \mathbb{R}^2 \times (\mathbb{R}^*)^2$, we say that X follows the one dimensional $MN(\phi, \mu_1, \mu_2, \sigma_1, \sigma_2)$ distribution if its density function is a convex combination of two Gaussian density functions: $\forall x \in \mathbb{R}$,

$$d_{MN}(x) = \frac{\phi}{\sqrt{2\pi\sigma_1^2}}e^{-\frac{(x-\mu_1)^2}{2\sigma_1^2}} + \frac{1-\phi}{\sqrt{2\pi\sigma_2^2}}e^{-\frac{(x-\mu_2)^2}{2\sigma_2^2}}. \tag{2.43}$$

The associated moment generating function may be easily derived from the Gaussian one: $\forall u \in \mathbb{R}$,

$$\mathbb{G}_{MN}(u) = \phi e^{u\mu_1 + \frac{\sigma_1^2u^2}{2}} + (1-\phi)e^{u\mu_2 + \frac{\sigma_2^2u^2}{2}} \tag{2.44}$$

from which we deduce its mean, variance, skewness and kurtosis

$$E[X] = \phi\mu_1 + (1-\phi)\mu_2$$

$$Var[X] = \phi\mu_1^2 + (1-\phi)\mu_2^2 + \phi\sigma_1^2 + (1-\phi)\sigma_2^2 - E[X]^2$$

$$sk[X] = \frac{1}{Var[X]^{\frac{3}{2}}} \sum_{i=1}^{2} \phi_i(\mu_i - E[X])(3\sigma_i^2 + (\mu_i - E[X])^2)$$

$$k[X] = \frac{1}{Var[X]^2} \sum_{i=1}^{2} \phi_i(3\sigma_i^4 + 6(\mu_i - E[X])^2\sigma_i^2 + (\mu_i - E[X])^4)$$

where $\phi_1 = \phi$ and $\phi_2 = 1 - \phi$. Using the preceding formulas, it is proved in Bertholon et al. (2006) that any possible pair of skewness-kurtosis can be reached by a mixture of only two Gaussian distributions. Moreover, as in the GH case, this distribution is stable with respect to affine transformations. If $X \hookrightarrow MN(\phi, \mu_1, \mu_2, \sigma_1, \sigma_2)$, then, $\forall (M, \Sigma) \in \mathbb{R} \times \mathbb{R}^*$,

$$M + \Sigma X \hookrightarrow MN(\phi, M + \Sigma\mu_1, M + \Sigma\mu_2, \Sigma\sigma_1, \Sigma\sigma_2).$$

Nevertheless, the MN distribution fulfills several analytic properties that makes it intrinsically different from the GH one. In fact, without any restriction on the parameters, it is stable by convolution (contrary to the GH family where the NIG distribution is the only one to have such property) making possible the explicit computation of conditional distributions of aggregated log-returns at any horizon. Even if it is not infinitely divisible (see Kelker 1971), and so more adapted to discrete time modeling than continuous one, this distribution allows for bimodal densities (see Behboodian 1970 and Fig. 2.12) and has an intuitive economic

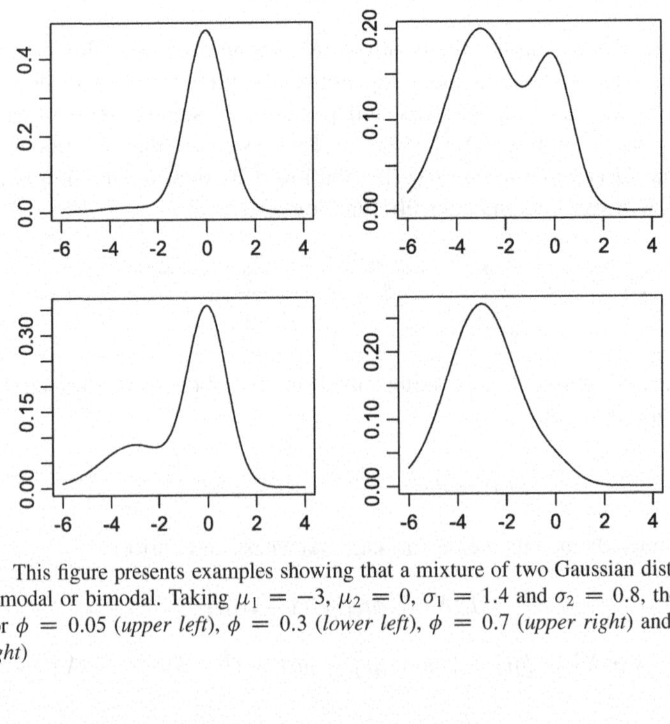

Fig. 2.12 This figure presents examples showing that a mixture of two Gaussian distributions is either unimodal or bimodal. Taking $\mu_1 = -3$, $\mu_2 = 0$, $\sigma_1 = 1.4$ and $\sigma_2 = 0.8$, the density is plotted for $\phi = 0.05$ (*upper left*), $\phi = 0.3$ (*lower left*), $\phi = 0.7$ (*upper right*) and $\phi = 0.95$ (*lower right*)

interpretation in the sense that the components represent different states of the market (e.g. market periods with different levels of volatility).

Due to its flexibility in accommodating any shape of continuous distributions, the MN distribution have been considered as a natural candidate in empirical finance (Kon 1984; Venkataraman 1997) in particular when it is combined with more or less refined GARCH structures (see Bai et al. 2004; Haas et al. 2004; Alexander and Lazar 2006; Bertholon et al. 2006; Badescu et al. 2008). We will test its empirical pricing performances in Chap. 4.

2.4.3 Some Practical Remarks

Three particular empirical and theoretical features mainly led to the popularity of the GH and MN distributions for financial applications and make them a natural choice to drive the GARCH-type models used in the Chap. 3 for pricing financial derivatives:

1. They are compatible with non zero skewness and large kurtosis observed in most of financial datasets and offer a better fit than the Gaussian distribution (see Fig. 2.13).
2. They are stable under affine transformations allowing, in the GARCH setting, to deduce the conditional distribution of the log-returns from the innovations. In particular they are perfectly adapted to Girsanov-type approaches of option pricing based on shift arguments (see Sect. 3.3).
3. They have an explicit and simple moment generating function. This is a necessary condition to be able to use the conditional Esscher transform to select an equivalent martingale measure in the discrete time setting (see Sect. 3.4).

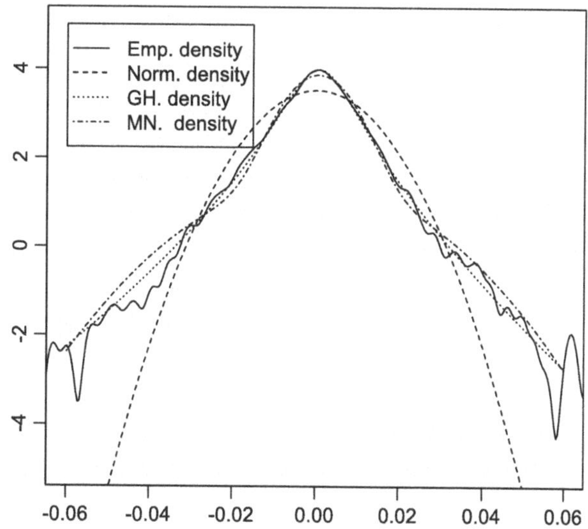

Fig. 2.13 This figure presents the empirical log-density vs. the estimated log-densities obtained with the Normal, GH and MN distributions using the S&P500 returns data set. Time varying variance has been filtered out using an EGARCH process to obtain the residuals. The sample starts on January 3, 1990 and ends on April 18, 2012

2.5 GARCH in Mean

For the preceding symmetric or asymmetric GARCH specifications we have considered, the conditional mean is by construction equal to zero (we speak about pure GARCH processes). When (X_t) represents a financial time series of log-returns, this is compatible with low empirical mean values (see Table 2.2). Nevertheless, under this hypothesis, there are no particular relations between conditional expected returns and risk levels. Considering, for simplicity, the classical GARCH(1,1) model (2.9), it is easy to introduce a time-varying excess return m_t

$$\begin{cases} X_t = r + m_t + \underbrace{\sqrt{h_t} z_t}_{\varepsilon_t} \\ h_t = a_0 + a_1 \varepsilon_{t-1}^2 + b_1 h_{t-1} \end{cases} \tag{2.45}$$

where r is a nonnegative real number that represents a risk free rate supposed to be constant and m_t is $\mathcal{F}_{t-1}$ measurable. The computation of the two first conditional moments gives

$$E[X_t \mid \mathcal{F}_{t-1}] = r + m_t, \; E[X_t^2 \mid \mathcal{F}_{t-1}] = h_t + (r + m_t)^2$$

and the conditional variance remains unchanged explaining why this family is simply called GARCH in mean models. Note that in (2.45), the shock driving the changes in the variance equation is not X_{t-1}^2 but still ε_{t-1}^2. The choice of a particular form for m_t, that is a priori only restricted by measurability conditions, is fundamental not only for empirical issues but also to preserve the analytic properties of the GARCH(1,1) model. One major attempt in this direction was done by Engle et al. (1987) taking m_t as a function $g(h_t)$ of the conditional variance h_t. This choice is particularly relevant because, as in the intertemporal Capital Asset Pricing Model of Merton (1973), it allows the conditional variance to be related to the conditional mean of the process, i.e., the expected risk premium. Moreover, the conditions of existence of a strictly stationary solution X_t to (2.45) remain, in this case, unchanged with respect to the pure GARCH ones because X_t is strictly stationary as long as the variance process is.[4] There is no empirical consensus on the particular form of g that may be found in the literature proportional to $\sqrt{h_t}$, h_t or $log(h_t)$ (see Christensen et al. 2012). Following Duan (1995) and having option pricing in mind, we will consider in the empirical Sect. 4.3

$$m_t = \lambda_0 \sqrt{h_t} - \frac{h_t}{2} \tag{2.46}$$

[4]This property is not preserved if we include, for example, in the conditional mean ARMA terms (see Nelson 1991).

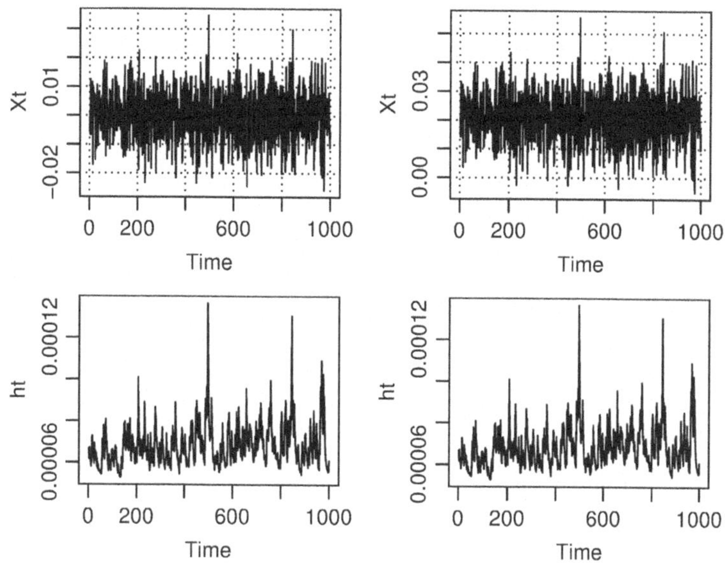

Fig. 2.14 Simulations of size 1,000 of a GARCH(1,1) process X_t and its associated conditional variance h_t with $a_0 = 10^{-5}$, $a_1 = 0.05$, $b_1 = 0.8$, $z_t \hookrightarrow \mathcal{N}(0,1)$ (*left column*) and of a GARCH(1,1) in mean process X_t with $r = 0.02$, $m_t = \lambda_0\sqrt{h_t} - \frac{h_t}{2}$ with $\lambda_0 = 0.1$ and its associated conditional variance h_t with $a_0 = 10^{-5}$, $a_1 = 0.05$, $b_1 = 0.8$ and $z_t \hookrightarrow \mathcal{N}(0,1)$ (*right column*). Both models are second order stationary

where λ_0 is the constant unit risk premium (see Fig. 2.14) that has to be estimated and that coincides with the discrete time counterpart of the so-called Black and Scholes (1973) market price of risk when the residuals are Gaussian and the conditional volatility constant.

2.6 Dealing with the Estimation Challenge

We consider a GARCH in mean model starting from $t = 0$: $\forall t \in \{1, \ldots, T\}$,

$$X_t = r + m_t(\lambda_0) + \sqrt{h_t} z_t, \quad z_0 = x \in \mathbb{R}, \tag{2.47}$$

$$h_t = F_{\theta^V}(z_{t-1}, h_{t-1}) \tag{2.48}$$

where the $(z_t)_{t\in\{1,\ldots,T\}}$ are independent and identically distributed real random variables, where $F_{\theta^V} : \mathbb{R}^2 \to \mathbb{R}_+$ is compatible with realistic GARCH(1,1)-type volatility models indexed by a set of parameters θ^V and where $m_t(\lambda_0)$ is a function of h_t depending on a constant unit risk premium λ_0 [see e.g. (2.46)]. From now on, we suppose that h_0 is a constant in such a way that the information filtration (generated by the (z_t)) is also generated by the log-returns (X_t). Concerning the innovations $(z_t)_{t\in\{1,\ldots,T\}}$ we additionally assume that their distribution lies in a parametric class of Lebesgue densities d_{θ^D} with mean 0 and variance 1 driven by a set of parameters θ^D. Starting from an observed data set $(x_1,\ldots,x_T)$, we discuss in this section several approaches to estimate the vector $(\lambda_0,\theta^V,\theta^D)$. The technical issues of these estimation strategies are clearly out of the scope of this book and we refer the reader to Straumann (2005) for a deep theoretical survey. The practical issues will be tackled in the empirical section.

2.6.1 Maximum Likelihood

Supposing that $(x_1,\ldots,x_T)$ is a realization of the model (2.47) and (2.48), it is possible to estimate the parameter $\theta = (\lambda_0,\theta^V,\theta^D)$ using a conditional version of the classical maximum likelihood (ML) estimation that is, in general, based on i.i.d observations. In fact, in this setting, the conditional log-likelihood based on the observations $(x_1,\ldots,x_T)$ is

$$L_T(x_1,\ldots,x_T \mid \theta) = \sum_{t=1}^{T} l_t(x_t \mid x_1,\ldots,x_{t-1},\theta) \tag{2.49}$$

where

$$l_t(x_t \mid x_1,\ldots,x_{t-1},\theta) = -\frac{log(h_t)}{2} + log\left[d_{\theta^D}\left(\frac{x_t - r - m_t(\lambda_0)}{\sqrt{h_t}}\right)\right]. \tag{2.50}$$

The ML estimator is defined, when it exists, by

$$\hat{\theta}_T = argmax_{\theta\in\Theta} L_T(x_1,\ldots,x_T \mid \theta) \tag{2.51}$$

or equivalently (when differentiability conditions are fulfilled) as the solution of the vectorial score equation

$$\sum_{t=1}^{T} \frac{\partial l_t(x_t \mid x_1,\ldots,x_{t-1},\theta)}{\partial\theta} = 0. \tag{2.52}$$

Remark 2.6.1 In order to implement the maximum likelihood algorithm, an explicit assumption regarding the conditional density has to be done. In the Gaussian case, the log-likelihood (2.49) simply becomes

$$L_T(x_1,\ldots,x_T \mid \theta) = \sum_{t=1}^{T} -\frac{log(h_t)}{2} - \frac{log(2\pi)}{2} - \frac{(x_t - r - m_t(\lambda_0))^2}{2h_t}. \tag{2.53}$$

We refer the reader to Straumann (2005, Chap. 6) for general conditions for the consistency and the asymptotic normality of this estimator. Under these conditions, the asymptotic variance-covariance matrix of the estimator is given by the inverse of the Fisher Information matrix

$$I(\theta) = -E\left[\frac{\partial^2 L_T(x_1,\ldots,x_T \mid \theta)}{\partial\theta^2}\right],$$

the expectation being taken under the true distribution. However and despite its efficiency this method is very sensitive to the misspecifications of the innovations' density. Newey and Steigerwald (1997) were the first to explain why the density used for the conditional distribution of the time series model has to be rich enough to avoid inconsistency problems. From a numerical point of view this leads to optimization issues that grow with the number of parameters, such as local maxima.

2.6.2 Quasi Maximum Likelihood

Since it overcomes the misspecification dilemma, the Quasi Maximum Likelihood (QML) estimator is probably the most famous estimation strategy for conditionally heteroscedastic time series. The likelihood (2.49) is written as if the distribution of the z_t was a Gaussian $\mathcal{N}(0,1)$:

$$L_T(x_1,\ldots,x_T \mid \lambda_0,\theta^V) = \sum_{t=1}^{T} -\frac{log(h_t)}{2} - \frac{log(2\pi)}{2} - \frac{(x_t - r - m_t(\lambda_0))^2}{2h_t} \tag{2.54}$$

and the QML estimator $(\widetilde{\lambda_{0,T}},\widetilde{\theta_T^V})$ is the argmax of (2.54). A remarkable feature is that this Gaussian assumption is not necessary in general to ensure good asymptotic properties of $(\widetilde{\lambda_{0,T}},\widetilde{\theta_T^V})$. In the case of pure GARCH processes, very mild conditions are necessary, as illustrated in Francq and Zakoian (2010, Chap. 7). However for more general GARCH specifications of the volatility only recent and partial theoretical results are available [see Straumann (2005, Chap. 5) for the pure AGARCH case, Meitz and Saikkonen (2012) for non-linear AR.GARCH models, Wintenberger (2013) for the pure EGARCH(1,1) model and Francq et al. (2013) for the pure log-GARCH process]. In a second step, the nuisance parameter θ^D is estimated from the previously obtained standardized residuals

$$\left(\frac{x_1 - r - m_1(\widetilde{\lambda_{0,T}})}{\sqrt{h_1}(\widetilde{\theta_T^V})},\ldots,\frac{x_T - r - m_T(\widetilde{\lambda_{0,T}})}{\sqrt{h_T}(\widetilde{\theta_T^V})}\right)$$

maximizing

$$\sum_{t=1}^{T} log\left[d_{\theta^D} \left(\frac{x_t - r - m_t(\widetilde{\lambda_{0,T}})}{\sqrt{h_t}(\widetilde{\theta_T^V})} \right) \right].$$

The main interest of this two-steps approach is to consider separately the distribution and volatility parameters, reducing in particular the dimension of the optimization problems, even if the estimators may become inefficient (see Francq and Zakoian 2010, Chap. 9).

2.6.3 Recursive Estimation

In our empirical experiments, we use another estimation approach introduced in Chorro et al. (2014) that specifically aims at dealing with the potential inefficiency of the two previously mentioned methodologies. Such efficiency issues are all the more likely in our experiments as asymmetric GARCH models mixed with asymmetric conditional distributions use the volatility and distribution set of parameters in order to span a single dimension: the potential dynamic skewness of the underlying time series. The empirical recursive method (REC) starts from the QML estimate and iteratively maximizes the likelihood over either the volatility or the distribution parameters:

1. We start from the QML estimate $\hat{\theta}_T^1 = (\widetilde{\lambda_{0,T}}, \widetilde{\theta_T^V}, \widetilde{\theta_T^D})$.
2. We re-estimate the conditional mean and volatility parameters (λ_0, θ^V) by maximizing

$$\sum_{t=1}^{T} -\frac{log(h_t)}{2} + log\left[d_{\widetilde{\theta_T^D}} \left(\frac{x_t - r - m_t(\lambda_0)}{\sqrt{h_t}} \right) \right]$$

   and we get $(\hat{\lambda}_{0,T}^2, \hat{\theta}_T^{2,V})$.
3. We re-estimate the distribution parameter θ^D from the standardized residuals

$$\left(\frac{x_1 - r - m_1(\hat{\lambda}_{0,T}^2)}{\sqrt{h_1}(\hat{\theta}_T^{2,V})}, \ldots, \frac{x_T - r - m_T(\hat{\lambda}_{0,T}^2)}{\sqrt{h_T}(\hat{\theta}_T^{2,V})} \right)$$

by maximizing

$$\sum_{t=1}^{T} log \left[d_{\theta^D} \left(\frac{x_t - r - m_t(\hat{\lambda}^2_{0,T})}{\sqrt{h_t}(\hat{\theta}^{2,V}_T)} \right) \right]$$

and obtain $\hat{\theta}^{2,D}_T$ and $\hat{\theta}^2_T = (\hat{\lambda}^2_{0,T}, \hat{\theta}^{2,D}_T, \hat{\theta}^{2,V}_T)$.

4. We iterate the two previous steps until a good trade off between precision and computational cost is attained.

The precision and computational costs of each of these three estimation methods is assessed numerically through naive Monte Carlo experiments in the next subsection.

2.6.4 Empirical Finite Sample Properties of the Three Estimation Methodologies

In this subsection, we review and detail the results of some Monte Carlo experiments developed to test the finite sample properties of the three estimation strategies presented earlier. The comparisons are based on simulated datasets coming from processes mixing a special time-varying volatility structure with a non-Gaussian distribution. For simplicity, we consider in this subsection only pure GARCH(1,1)-type models: using the notations of the preceding subsection, we take $r = m_t = 0$ in Eq. (2.47). In practice we use the APARCH and EGARCH volatility structures presented in Sect. 2.3 and the GH and MN distributions introduced in Sect. 2.4. The numerical features of the simulation strategy implemented here unfold as follows: the total number of experiments with each strategy for each model is 2,000 and the sample size used each time is 1,500 observations. The simulations are based on parameters selected so that to mimic the salient features of financial markets returns:

- For the APARCH volatility structure (2.38) we take $\Delta = 1.2$, $a_0 = 0.04$, $a_1 = 0.3$, $b_1 = 0.6$ and $\gamma = 0.75$,
- For the EGARCH (2.34) volatility structure we take $a_0 = -0.4$, $a_1 = 0.1$, $b_1 = 0.96$ and $\gamma = 1$,
- For the MN density (2.43), the selected parameters are $\phi = 0.23$, $\mu_1 = -0.4$, $\sigma_1 = 1.3$, $\mu_2 = 0.12$ and $\sigma_2 = 0.86$,
- For the GH distribution (2.39) the selected parameters are $\alpha = 1.9$, $\beta = -0.55$, $\delta = 3.6$, $\mu = 0.55$, $\lambda = -5.5$.

For a given model, we generate a sample of size 1,500 using the true value for the parameters. Then, we perturb[5] these true values to obtain a starting point to optimize the likelihood function using the previous sample and the three estimation methods.[6] At the end of this process, we store the relevant information and start again these steps 2,000 times. The REC estimation approach is run ten times: this number was selected on the ground of various information. First, the improvement over this tenth step was unclear during our initial trial-and-error. Second, the average time required to obtain these 10 optimization steps is around 6–9 times the time required for the QML estimator to be run. We wanted to make the approaches broadly comparable, even from a time consumption perspective. In order to compare the estimation approaches, we propose to rely on the two following criteria:

The first criterion that we rely on is the Root Mean Square Error (RMSE hereafter) of the estimated parameters. It is due to help us deciding upon which methodology yields estimated parameters that are the closest to the true values used to sample the process. Let θ_i^0 be the true value for the parameter i and $\hat{\theta}_i^{j,n}$ be the estimated parameter using the methodology j, obtained from the nth simulated sample. This parameter can belong either to the volatility structure or to the conditional distribution one. With a total number of $N = 2{,}000$ simulations and I parameters, the RMSE[7] for the methodology j is

$$RMSE_j = \sqrt{\frac{1}{N}\sum_{n=1}^{N}\sum_{i=1}^{I}\left(\frac{\theta_i^0 - \hat{\theta}_i^{j,n}}{\theta_i^0}\right)^2}. \tag{2.56}$$

In the end, we compute three different scores. The first one is the previous score computed over the total number of parameters that we refer to as the "Total RMSE".

[5]For each experiment, the starting values of the parameters are obtained by perturbing the true values stated above in the following way:

$$\theta_i^{j,n} = \theta_i^0\left(\frac{1}{2} + u\right), \tag{2.55}$$

for a given estimation approach j and the nth replication of the Monte Carlo experience. The number u is the realization of a random variable which follows a uniform distribution over $[0, 1]$. There are numbers of constraints required with these volatility structures and distributions. Once the parameters are perturbed, we check for their consistency with these constraints and discard those that do not match these requirements. What is more, we impose these constraints numerically within the optimization process. However, this is of little impact on the results as the starting point is selected to be close to the true value of the parameters. This would have a sharper influence on the results in the case of a real data set, involving the difficult step of the initialization of the parameters without knowing them.

[6]The volatility is initialized to its long term average as estimated from the sample using the method of moment estimator.

[7]Larger parameters are usually affected by larger estimation errors: in the criterion that we propose, the errors are weighted by the true value of the parameter, making the aggregation of the estimation errors coherent using relative quantities that are scale independent.

The second and third scores focus on specific parameters: we are interested in the differences obtained with each estimation strategy when it either comes to the volatility parameters or to the distribution ones. We thus compute the previous RMSE criterion for these two different subsets of parameters. We refer to them as "Volatility RMSE" and "Distribution RMSE".

Maximizing a likelihood function over parameters usually requires to use a numerical optimizer. Most of these optimizers are gradient-based methods and in our experiments we rely on the Broyden-Fletcher-Goldfarb-Shanno (BFGS) optimization algorithm. This method is a quasi-Newton method and interested readers can refer to Nocedal and Wright (1999). Such algorithm requires a certain number of iterations before it delivers the output of the likelihood maximization and takes time. The second numerical aspect we are interested in is thus the time used by each method: computational burden is an aspect of numerical optimization that should not be left apart, as the feasibility of the approaches investigated here should be an important trigger for readers interested in our results. Hence, the second criterion is the average time required to perform the estimation of each simulated model using the different estimation strategies presented before.

Now, we turn our attention to the detailed analysis of the results, specifying the outcomes of this simulation-based assessment of the various estimation approaches:

- **Case 1: the GH-APARCH model (Table 2.7)**

For this case, the lowest Total RMSE is obtained with the final step of the recursive estimation approach, whose value is 2.66. It improves the QML starting point whose value is 3.65. The ML approach behaves poorly, yielding a Total RMSE that is equal to 4.69. The REC10 improvement stems mainly from the estimation of the volatility parameters, as the ML approach is the worst performing competitor, with a Volatility RMSE equal to 3.99, vs. 3.45 for the QML and 2.54 for the REC10.

Table 2.7 Comparison of the estimation methodologies in the case of the GH-APARCH model

Method	Time	Distribution RMSE	Volatility RMSE	Total RMSE
ML	12.02	0.64	3.99	4.04
QML	6.16	1.21	3.45	3.65
REC2	21.81	2.08	11.47	11.66
REC3	27.1	2.23	10.62	10.86
REC4	31.9	2.15	9.51	9.75
REC5	36.28	1.98	8.26	8.49
REC6	40.18	1.66	6.9	7.09
REC7	43.63	1.53	5.89	6.08
REC8	47.08	1.19	4.29	4.45
REC9	50.77	1.15	3.91	4.08
REC10	54.49	0.8	2.54	2.66

Table 2.8 Comparison of the estimation methodologies in the case of the MN-APARCH model

Method	Time	Distribution RMSE	Volatility RMSE	Total RMSE
ML	13.99	0.43	4.67	4.69
QML	6.67	0.75	2.99	3.08
REC2	25.43	1.03	10.81	10.86
REC3	29.68	1.02	9.08	9.14
REC4	34.05	0.94	7.91	7.97
REC5	38.18	0.84	6.47	6.52
REC6	42.32	0.75	5.25	5.3
REC7	46.46	0.68	4.06	4.12
REC8	50.87	0.62	3.4	3.45
REC9	55.55	0.58	2.67	2.73
REC10	60.54	0.55	2.09	2.16

When it comes to the distribution's parameters estimation, the ML becomes the best competitor with a RMSE of 0.64, vs. 1.21 for the QML and 0.8 for REC10. In the end, the improvement over the volatility structure estimation is so important with the REC10 approach that it dominates the Total RMSE obtained. Finally, the average time required to perform the estimation is 6 s for QML, 12 s for ML and 55 s for REC10.[8]

– **Case 2: the MN-APARCH model (Table 2.8)**

This case is very similar to the latter as the REC10 approach delivers the best Total RMSE (2.16), improving not only the QML score (3.08), but also the ML case (4.69). Again, the main explanation stems from the volatility's parameters estimation as for the Volatility RMSE, the hierarchy between the approaches remains the same: ML obtains 4.67, QML 2.99 and REC10 2.09. The distribution's parameters are best estimated with the ML method (0.43), whereas the REC10 one comes second (0.55) while still improving the QML case (0.75). However, these figures remain remarkably close to each other. From a convergence time REC10 uses 61 s when the ML estimators are obtained after 14 s and the QML ones after 7 s.

– **Case 3: the GH-EGARCH model (Table 2.9)**

The results obtained here are different from those obtained before: it appears that QML gets the lowest Total RMSE (3.12), while being very close to the ML estimates (3.14). The REC10 delivers a score of 3.64, which remains however close to the others. This is again the result of the estimation of the volatility parameters,

[8]This exercise has been done using a Intel Xeon E5420 PC (2.5 GHz, 1,333 MHz, 2X6 Mo).

Table 2.9 Comparison of the estimation methodologies in the case of the GH-EGARCH model

Method	Time	Distribution RMSE	Volatility RMSE	Total RMSE
ML	9.82	0.76	3.04	3.14
QML	5.97	0.99	2.96	3.12
REC2	14.34	1.15	5.12	5.25
REC3	26.32	1.01	8.26	8.32
REC4	33.56	0.91	6.13	6.2
REC5	47.04	0.85	7.93	7.98
REC6	49.47	0.77	9.43	9.47
REC7	52.05	0.73	7.35	7.38
REC8	54.95	0.7	7.37	7.41
REC9	58.09	0.68	4.64	4.69
REC10	62.14	0.66	3.58	3.64

Table 2.10 Comparison of the estimation methodologies in the case of the MN-EGARCH model

Method	Time	Distribution RMSE	Volatility RMSE	Total RMSE
ML	7.91	0.55	1.68	1.77
QML	7.15	0.92	3.02	3.16
REC2	8.5	3.4	3.14	4.63
REC3	14.05	3.26	10.44	10.94
REC4	19.38	2.63	10.51	10.84
REC5	21.27	2.3	10.24	10.49
REC6	25.91	1.96	8.72	8.94
REC7	28.85	1.31	8.73	8.83
REC8	33.26	1.13	8.64	8.71
REC9	37.67	0.94	1.36	1.65
REC10	42.03	0.76	1.15	1.38

as a similar ranking is obtained with the Volatility RMSE: QML yields 2.96, ML gets 3.04 and finally REC10 delivers a ratio of 3.58. Here again the scores remain close. Finally, the QML approach is the fastest strategy, with an average of 5.97 s, when ML yields estimates after 9.82 s and REC10 after 62.14 s.

- **Case 4: the MN-EGARCH model (Table 2.10)**

For such a case, the REC10 dominates again the Total RMSE ranking (1.38), whereas QML is the worst performing competitor (3.16), right behind ML (1.77). In this case, the volatility results explain most of this performance, as the REC10 shrinks the QML's Volatility RMSE from 3.02 to 1.15. ML gets a score of 1.68. The story for the Distribution RMSE is different again, as ML appears to be the best competitor with a score of 0.55, when QML gets a score of 0.92 that is reduced by the REC10 to 0.76. The gap between ML and REC10 is thus small. This result is thus similar to the one obtained in the GH-EGARCH. The time consumption in this case for the REC10 approach is smallest of all our experiments, as it requires only 42 s. The same comment applies for the ML whose average time needed to estimate the parameters is 8 s.

We can thus summarize the salient features of this exercise as follows: first, the REC10 approach always improves the QML one, whatever the model and the subset of parameters investigated. Second, the REC10 dominates 3 out of the 4 Total RMSE investigated, mostly because it provides estimates of the volatility parameters with smaller estimation errors. The results obtained in the case of the distribution's parameters do not provide such sharp conclusions. Even though the ML estimates remain the most precise ones, the difference between the REC10 estimates and the ML ones remains small in every case. A notable exception is obtained in the GH-EGARCH case, with a lower Distribution RMSE in the REC10 case than in the ML one. This overall nice performance of the REC10 approach comes at the cost of a longer estimation time. In Sect. 4.3.1, the intuitions gathered in these Monte Carlo experiments are completed by a deep empirical study based on S&P500 log-returns.

Before to conclude this chapter, we highlight in the next subsection the relations that may exist between GARCH and continuous time processes, two areas developed along very separate lines until the 1990s and the seminal work of Nelson (1990b).

2.7 From GARCH Processes to Continuous Diffusions

Continuous time stochastic volatility models are commonly employed in financial analysis and most of theoretical results on derivative pricing and hedging rely on the continuous time hypothesis. This importance is partly explained by the mathematical tractability of these models even if estimation of the parameters solely using the historical time series of returns is often tricky. Popular models include bivariate diffusions as presented in Hull and White (1987) or Heston (1993). In this section we investigate the continuous time limit of some Gaussian and non-Gaussian GARCH models presented earlier. The aim is not to provide theoretical general convergence results but to make the reader familiar with the intriguing link between these two classes of models by studying well-chosen examples.

Nelson (1990b) was the first study that derives a diffusion limit for GARCH processes as the length of the stepsize between two trading dates shrinks to zero [see also Duan (1997) and Badescu et al. (2013) for related extensions]. The obtained results rely on notions introduced in Stroock and Varadhan (1979) to study the convergence of Markov processes to continuous diffusions (see Theorem A.3 of the Appendix) and are generally used in the literature to justify the similarities between GARCH and diffusion models. Nevertheless, from the work of Corradi (2000), it is now well documented that the convergence of GARCH processes deeply depends on the parametric constraints on the time bucket impact on parameters and that analogies have to be done carefully.[9] In the three next examples, we see that diffusion limits may lead to very different noise propagations.

2.7.1 Convergence Toward Hull and White (1987) Diffusions

For $T \in \mathbb{R}_+^*$ and $n \in \mathbb{N}^*$, we consider a GARCH in mean process $(X_{k\tau}^{(n)})_{k\in\{0,\dots,nT\}}$ indexed by the time unit $\tau = \frac{1}{n}$ with a conditional mean structure given by (2.46). If we denote by $(Y_{k\tau}^{(n)} = log(S_{k\tau}^{(n)}))_{k\in\{0,\dots,nT\}}$ the logarithm of the risky asset price process, we have the following stochastic volatility structure[10]:

$$\begin{cases} X_{k\tau}^{(n)} = Y_{k\tau}^{(n)} - Y_{(k-1)\tau}^{(n)} = \left(r + \lambda_0\sqrt{h_{k\tau}^{(n)}} - \frac{h_{k\tau}^{(n)}}{2}\right)\tau + \sqrt{\tau}\underbrace{\sqrt{h_{k\tau}^{(n)}}z_{k\tau}^{(n)}}_{\varepsilon_{k\tau}^{(n)}}, \\ h_{k\tau}^{(n)} = a_0(\tau) + a_1(\tau)\left(\varepsilon_{(k-1)\tau}^{(n)}\right)^2 + b_1(\tau)h_{(k-1)\tau}^{(n)} \end{cases} \tag{2.57}$$

where the $(z_{k\tau}^{(n)})_{k\in\{0,\dots,nT\}}$ are i.i.d random variables with a finite fourth order moment such that

$$E[z_{k\tau}^{(n)}] = 0,\ Var[z_{k\tau}^{(n)}] = 1,\ E[(z_{k\tau}^{(n)})^3] = \mu_3 \text{ and } E[(z_{k\tau}^{(n)})^4] = \mu_4.$$

We suppose that $(Y_0^{(n)}, h_0^{(n)}) = (y_0, h_0) \in \mathbb{R}\times\mathbb{R}_+^*$ and that $(a_0(\tau), a_1(\tau), b_1(\tau))$ are nonnegative constants depending on τ such that $a_0(\tau) > 0$ and $a_1(\tau) + b_1(\tau) < 1$.

We remark that $Z_{k\tau}^{(n)} = (Y_{k\tau}^{(n)}, h_{(k+1)\tau}^{(n)})$ is a Markov chain that may be embedded into a continuous time process $(Z_t^{(n)})_{t\in[0,T]}$ by defining

$$Z_t^{(n)} = Z_{k\tau}^{(n)} \text{ if } k\tau \le t < (k+1)\tau.$$

We obtain after some tedious computations that

$$\frac{1}{\tau}E\left[Z_{(k+1)\tau}^{(n)} - Z_{k\tau}^{(n)} | Z_{k\tau}^{(n)} = (y,h)\right] = \left(r + \lambda_0\sqrt{h} - \frac{h}{2}, \frac{a_0(\tau) + (a_1(\tau) + b_1(\tau) - 1)h}{\tau}\right), \tag{2.58}$$

---

[9]We can also mention on this topic, the important result of Wang (2002) that shows that statistical inference for GARCH modelling and statistical inference for the diffusion limit are not equivalent in general.

[10]The factor $\sqrt{\tau}$ (resp. τ) that appears in the conditional variance (resp. conditional mean) of the log-returns $X^{(n)}$ over a time period of length τ is introduced to respect classical formulas to convert volatility (resp. conditional mean) from one time period to another. Note that in the special case where $\tau = 1$, the model (2.57) reduces to the classical GARCH in mean process (2.45).

$$\frac{1}{\tau}Var\left[Z^{(n)}_{(k+1)\tau}-Z^{(n)}_{k\tau}|Z^{(n)}_{k\tau}=(y,h)\right]=\begin{pmatrix} h & \frac{a_1(\tau)}{\sqrt{\tau}}h^{\frac{3}{2}}\mu_3 \\ \frac{a_1(\tau)}{\sqrt{\tau}}h^{\frac{3}{2}}\mu_3 & \frac{a_1(\tau)^2}{\tau}h^2(\mu_4-1)\end{pmatrix}. \tag{2.59}$$

If we suppose that

$$\lim_{\tau\to 0}\frac{a_0(\tau)}{\tau}=w_0,\ \lim_{\tau\to 0}\frac{a_1(\tau)+b_1(\tau)-1}{\tau}=-w_1 \text{ and } \lim_{\tau\to 0}\frac{a_1(\tau)^2}{\tau}=w_2 \text{[11]} \tag{2.60}$$

we get

$$\lim_{\tau\to 0}\frac{1}{\tau}E\left[Z^{(n)}_{(k+1)\tau}-Z^{(n)}_{k\tau}|Z^{(n)}_{k\tau}=(y,h)\right]=\left(r+\lambda_0\sqrt{h}-\frac{h}{2},w_0-w_1h\right),$$

and

$$\lim_{\tau\to 0}\frac{1}{\tau}Var\left[Z^{(n)}_{(k+1)\tau}-Z^{(n)}_{k\tau}|Z^{(n)}_{k\tau}=(y,h)\right]=\begin{pmatrix} h & \sqrt{w_2}h^{\frac{3}{2}}\mu_3 \\ \sqrt{w_2}h^{\frac{3}{2}}\mu_3 & w_2h^2(\mu_4-1)\end{pmatrix}.$$

Remarking that

$$\begin{pmatrix} h & \sqrt{w_2}h^{\frac{3}{2}}\mu_3 \\ \sqrt{w_2}h^{\frac{3}{2}}\mu_3 & w_2h^2(\mu_4-1)\end{pmatrix}=\begin{pmatrix} \sqrt{h} & 0 \\ \sqrt{w_2}h\mu_3 & \sqrt{w_2}h\sqrt{\mu_4-1-\mu_3^2}\end{pmatrix}\begin{pmatrix} \sqrt{h} & \sqrt{w_2}h\mu_3 \\ 0 & \sqrt{w_2}h\sqrt{\mu_4-1-\mu_3^2}\end{pmatrix}$$

and using Theorem A.3 of the Appendix, we show the weak convergence of $(Z_t^{(n)})_{t\in[0,T]}$ toward the stochastic process $(Z_t=(Y_t,h_t))_{t\in[0,T]}$ that fulfills a Hull and White (1987) type stochastic differential equation:

$$\begin{cases} dY_t=\left(r+\lambda_0\sqrt{h_t}-\frac{h_t}{2}\right)dt+\sqrt{h_t}dW_t^1, \\ dh_t=(w_0-w_1h_t)dt+\sqrt{w_2}\mu_3h_tdW_t^1+\sqrt{w_2}\sqrt{\mu_4-1-\mu_3^2}h_tdW_t^2 \end{cases} \tag{2.61}$$

where W^1 and W^2 are two independent Brownian motions.[12]

---

[11]These parametric constraints are obtained if we take $a_0(\tau)=w_0\tau, a_1(\tau)=\sqrt{w_2\tau}$ and $b_1(\tau)=1-a_1(\tau)-w_1\tau$.

[12]This diffusion limit has been extended in Badescu et al. (2013) for the AGARCH case. When the residuals are Gaussian ($\mu_3=0$ and $\mu_4=3$) several extensions for classical asymmetric GARCH specifications may be found in Duan (1997). Let us also mention that, in the spirit of Duan et al. (2006), it is possible to consider GARCH extensions (called GARCH-Jump processes) that include, as limiting cases, processes characterized by jumps in both prices and volatilities.

2.7.2 Convergence Toward Diffusions with Deterministic Volatilities

In the previous framework, if we replace (see Corradi 2000) the parametric constraints (2.60) by

$$\lim_{\tau\to 0} \frac{a_0(\tau)}{\tau} = w_0, \; \lim_{\tau\to 0} \frac{a_1(\tau)+b_1(\tau)-1}{\tau} = -w_1 \text{ and } \lim_{\tau\to 0} \frac{a_1(\tau)}{\sqrt{\tau}} = 0, \qquad (2.62)$$

the volatility process in (2.61) becomes purely deterministic and fulfills the following ordinary differential equation

$$dh_t = (w_0 - w_1 h_t)dt.$$

As emphasized by this result, the strong dependence of diffusion limits of GARCH models on parametric constraints should prevent the reader to overhasty analogies. It also stresses the specific features of GARCH-type modeling with respect to the continuous time approach.

In the next example, we see an explicit GARCH-type parameterization that converges toward the so-called Heston (1993) model. Remark that this model is not a nested case of (2.61).

2.7.3 Convergence Toward the Heston (1993) Model

Taking the notations introduced in the first example, we assume, following Heston and Nandi (2000), that

$$\begin{cases} X_{k\tau}^{(n)} = Y_{k\tau}^{(n)} - Y_{(k-1)\tau}^{(n)} = \left(r + \lambda_0 h_{k\tau}^{(n)}\right)\tau + \sqrt{\tau}\sqrt{h_{k\tau}^{(n)}} z_{k\tau}^{(n)}, \\ h_{k\tau}^{(n)} = a_0(\tau) + a_1(\tau)\left(z_{(k-1)\tau}^{(n)} + \gamma(\tau)\sqrt{h_{(k-1)\tau}^{(n)}}\right)^2 + b_1(\tau) h_{(k-1)\tau}^{(n)} \end{cases} \qquad (2.63)$$

where the $(z_{k\tau}^{(n)})_{k\in\{0,\ldots,nT\}}$ are i.i.d standard Gaussian random variables.[13] Supposing that

$$a_0(\tau) = (\kappa\theta - \frac{\sigma^2}{4})\tau, \; a_1(\tau) = \frac{\sigma^2\tau}{4}, \; b_1(\tau) = 0 \text{ and } \gamma(\tau) = \frac{2}{\sigma\sqrt{\tau}} - \frac{\kappa\sqrt{\tau}}{\sigma}, \qquad (2.64)$$

[13]One of the main features of the Heston and Nandi (2000) model is that both conditional expectation and variance of the volatility process are affine functions of the volatility at the preceding trading date (we speak about affine models). This is not true for the classical GARCH model (2.53) where the conditional variance of the volatility has a quadratic form. This major difference explains not only why we pass from two to one Brownian motions in (2.61) and (2.65) but also why the price of vanilla options in the Heston–Nandi model has a pseudo analytic form (see Sect. 3.7).

we show (using once again Theorem A.3 of the Appendix) the weak convergence of $(Z_t^{(n)})_{t\in[0,T]}$ toward the stochastic process $(Z_t = (Y_t, h_t))_{t\in[0,T]}$ that fulfills the Heston (1993) stochastic differential equation involving a single Brownian motion W^1 :

$$\begin{cases} dY_t = (r + \lambda_0 h_t)\, dt + \sqrt{h_t} dW_t^1, \\ dh_t = \kappa(\theta - h_t)dt + \sigma \sqrt{h_t} dW_t^1. \end{cases} \tag{2.65}$$

To conclude, let us remark that for the bivariate diffusions that appear in this subsection as limits of GARCH-type models, the pricing of financial derivatives is now well documented and generally based on extensions of the Girsanov theorem to obtain equivalent martingale measures. The aim of the next chapter is to provide the reader with equivalent tools in the framework of the discrete time GARCH modeling.

References

Abramowitz M, Stegun I (1964) Handbook of mathematical functions with formulas, graphs and mathematical tables. Dover, New York

Alexander C, Lazar E (2006) Normal mixture GARCH(1,1): applications to exchange rate modelling. J Appl Econ 21:307–336

Andersen TG, Bollerslev T, Diebold X (2010) Parametric and nonparametric volatility measurement, chapter 2. In: Hansen LP, Aït-Sahalia Y (eds) Handbook of financial econometrics. Elsevier, Amsterdam, pp 67–128

Bachelier L (1900) Théorie de la spéculation. Ann Sci Ecole Norm Supér 17:21–86

Badescu A, Kulperger R, Lazar E (2008) Option valuation with normal mixture GARCH. Stud Nonlinear Dyn Econ 12(2):1–42

Badescu A, Elliott RJ, Kulperger R, Miettinen J, Siu TK (2011) A comparison of pricing kernels for GARCH option pricing with generalized hyperbolic distributions. Int J Theor Appl Finance 14(5):669–708

Badescu A, Elliott RJ, Ortega JP (2013) On the extended Girsanov principle for GARCH models and their diffusion limits. Working Paper

Bai X, Russell JR, Tiao GC (2004) Kurtosis of garch and stochastic volatility models with non-normal innovations. J Econ 114:349–360

Baillie RT, Bollerslev T (1992) Prediction in dynamic models with time dependent conditional variances. J Econ 52:91–113

Barndorff-Nielsen OE (1977) Exponentially decreasing distributions for the logarithm of particle size. Proc R Soc Lond Ser A 353:401–419

Barndorff-Nielsen OE (1978) Hyperbolic distributions and distributions on hyperbolae. Scand J Stat 5:151–157

Barndorff-Nielsen OE (1995) Normal inverse Gaussian processes and the modelling of stock returns. Research report 300. Department Theoretical Statistics, Aarhus University, Aarhus

Barndorff-Nielsen OE, Blaesild P (1981) Hyperbolic distributions and ramifications: contributions to theory and applications. In: Taillie C, Patil GP, Baldessari BA (eds) Statistical distributions in scientific work, vol 4. Reidel, Dordrecht, pp 19–44

Barndorff-Nielsen OE, Halgreen O (1977) Infinite divisibility of the hyperbolic and generalized inverse Gaussian distributions. Z Wahrscheinlichkeit Verwandte Gebiete 38:309–312

Behboodian J (1970) On the modes of a mixture of two normal distributions. Technometrics 12(1):131–139

Bertholon H, Monfort A, Pegoraro F (2006) Pricing and inference with mixtures of conditionally normal processes. Working Paper. CREST, Paris

Black F (1976) Studies of stock prices volatility changes. In: Proceedings of the 1976 meetings of the American Statistical Association, Business and Economics Statistics Section, pp 177–189

Black F, Scholes M (1973) The pricing of options and corporate liabilities. J Polit Econ 81:637–659

Blaesild P (1999) Generalized hyperbolic and generalized inverse Gaussian distributions. Working Paper. University of Arhus, Arhus

Bollerslev T (1986) Generalized autoregressive conditional heteroscedasticity. J Econ 31:307–327

Bollerslev T (1987) A conditionally heteroskedastic time series model for speculative prices and rates of return. Rev Econ Stat 69(2):542–547

Bollerslev T (2011) Glossary to ARCH (GARCH). In: Bollerslev T, Russell R, Watson M (eds) Volatility and time series econometrics: essays in honor of Robert F. Engle. Oxford University Press, Oxford

Bollerslev T, Chou RY, Kroner KF (1992) ARCH modeling in finance: a review of the theory and empirical evidence. J Econ 52:5–59

Bouchaud JP, Potters M (2000) Theory of financial risks. Cambridge University Press, Cambridge

Bougerol P, Picard N (1992) Stationarity of GARCH processes and some nonnegative time series. J Econ 52:115–127

Box GEP, Jenkins GM, Reinsel GC (2008) Time series analysis: forecasting and control. Wiley, New York

Brockwell P, Davis RA (1996) Time series: theory and methods. Springer, New York

Chorro C, Guégan D, Ielpo F (2012) Option pricing for GARCH type models with generalized hyperbolic innovations. Quant Finance 12(7):1079–1094

Chorro C, Guégan D, Ielpo F, Lalaharison H (2014) Testing for leverage effect in financial returns. CES Working Papers, 2014.22

Christensen BJ, Dahl CM, Iglesias EM (2012) Semiparametric inference in a GARCH-in-mean model. J Econ 167(2):458–472

Christoffersen P, Jacobs K (2004) Which garch model for option valuation? Manag Sci 50(9):1204–1221

Cont R (2001) Empirical properties of asset returns: stylized facts and statistical issues. Quant Finance 1:223–236

Cont R, Tankov P (2003) Financial modelling with jump processes. Chapman & Hall, London

Corradi V (2000) Reconsidering the continuous time limit of the GARCH(1,1) process. J Econ 96:145–153

Dacorogna M, Gençay R, Müller U, Olsen R, Pictet O (2001) An introduction to high-frequency finance. Academic, San Diego

Davis RA, Mikosch T (1998) Limit theory for the sample ACF of stationary process with heavy tails with applications to ARCH. Ann Stat 26:2049–2080

Ding Z, Granger CWJ, Engle RF (1993) A long memory property of stock market returns and a new model. J Empir Finance 1:83–106

Duan JC (1995) The GARCH option pricing model. Math Finance 5:13–32

Duan JC (1997) Augmented GARCH(p,q) process and its diffusion limit. J Econ 79:97–127

Duan JC, Ritchken P, Sun Z (2006) Approximating GARCH-jump models, jump diffusion processes, and option pricing. Math Finance 16:21–52

Eberlein E, Hammerstein EAV (2003) Generalized hyperbolic and inverse Gaussian distributions: limiting cases and approximation of processes. In: Dalang RC, Dozzi M, Russo F (eds) Proceedings of the 4th Ascona conference. Birkhauser, Basel

Eberlein E, Keller U (1995) Hyperbolic distributions in finance. Bernoulli 1:281–299

Eberlein E, Prause K (2002) The generalized hyperbolic model: financial derivatives and risk measures. In: Geman H, Madan D, Pliska S, Vorst T (eds) Mathematical finance-bachelier congress 2000. Springer, New York, pp 245–267

Embrechts P, Frey R, McNeil AJ (2005) Quantitative risk management. Princeton University Press, Princeton

Engle RF (1982) Autoregressive conditional heteroscedasticity with estimates of the variance of United Kingdom inflation. Econometrica 50:987–1007

Engle RF (1990) Discussion: stock market volatility and the crash of 87. Rev Financ Stud 3:103–106

Engle RF (2001) GARCH 101: an introduction to the use of arch/garch models in applied econometrics. NYU Working Paper No. FIN-01-030

Engle RF, Bollerslev T (1986) Modelling the persistence of conditional variances. Econ Rev 94:405–420

Engle RF, Ng VK (1993) Measuring and testing the impact of news on volatility. J Finance 48:1749–1777

Engle RF, Lilien DM, Robins RP (1987) Estimating time varying risk premia in the term structure: the ARCH-M model. Econometrica 55(2):391–407

Fama E (1965) The behavior of stock markets prices. J Bus 38:34–105

Fornari F, Mele A (1996) Modeling the changing asymmetry of conditional variances. Econ Lett 50:197–203

Francq C, Zakoian JM (2006) Mixing properties of a general class of GARCH(1,1) models without moment assumptions. Econ Theory 22:815–834

Francq C, Zakoian JM (2010) GARCH models: structure, statistical inference and financial applications. Wiley, New York

Francq C, Wintenberger O, Zakoian JM (2013) GARCH models without positivity constraints: exponential or Log GARCH? J Econ 177:34–46

Glosten LR, Jagannathan R, Runkle DE (1993) On the relation between the expected value and the volatility of the nominal excess return on stocks. J Finance 48(5):1779–1801

Gonzalez-Rivera G (1998) Smooth transition GARCH models. Stud Nonlinear Dyn Econ 3:161–178

Gouriéroux C (1997) ARCH models and financial applications. Springer, New York

Granger CWJ (2005) The past and future of empirical finance: some personal comments. J Econ 129:35–40

Granger CWJ, Andersen AP (1978) An introduction to bilinear time series analysis. Vandenhoeck and Ruprecht, Gottingen

Haas M, Mittnik S, Paolella M (2004) Mixed normal conditional heteroskedasticity. J Financ Econ 2:211–250

Hansen BE (1994) Autoregressive conditional density estimation. Int Econ Rev 35(3):705–730

He C, Terasvirta T (1999a) Fourth-moment structure of the GARCH(p,q) process. Econ Theory 15:824–846

He C, Terasvirta T (1999b) Properties of moments of a family of GARCH processes. J Econ 92:173–192

He C, Terasvirta T (1999c) Statistical properties of the asymmetric power ARCH process. In: Engle RF, White H (eds) Cointegration, causality and forecasting. A festschrift in honour of Clive W.J. Granger. Oxford University Press, New York

He C, Terasvirta T, Malmsten H (2002) Moment structure of a family of first-order exponential GARCH models. Econ Theory 18:868–885

Hentschel L (1995) All in the family. Nesting symmetric and asymmetric GARCH models. J Financ Econ 39:71–104

Heston SL (1993) A closed-form solution for options with stochastic volatility with applications to bond and currency options. Rev Financ Stud 6(2):327–343

Heston SL, Nandi S (2000) A closed-form GARCH option valuation. Rev Financ Stud 13:585–625

Hull J, White A (1987) The pricing of options on assets with stochastic volatilities. J Finance 42:281–300

Jensen MB, Lunde A (2001) The NIG-ARCH model: a fat tailed stochastic and autoregressive heteroscedastic volatility model. Econ J 4:319–342

Karanasos M, Kim J (2006) A re-examination of the asymmetric power ARCH model. J Empir Finance 13:113–128
Kelker D (1971) Infinite divisibility and variance mixtures of the normal distribution. Ann Math Stat 42(2):802–808
Kluppelberg C, Lindner A, Maller R (2004) A continuous-time GARCH process driven by a Lévy process: stationarity and second-order behavior. J Appl Probab 41(3):601–622
Kon SJ (1984) Models of stock returns: a comparison. J Finance 39:147–165
Ling S, McAleer M (2002a) Necessary and sufficient moment conditions for the GARCH(r,s) and the asymmetric power GARCH(r,s) models. Econ Theory 18:722–729
Ling S, McAleer M (2002b) Stationarity and the existence of moments of a family of GARCH processes. J Econ 106:109–117
Madan DB, Carr PP, Chang EC (1998) The variance gamma process and option pricing. Eur Finan Rev 2:79–105
Malmsten H, Terasvirta T (2004) Stylized facts of financial time series and three popular models of volatility. SSE/EFI Working Paper series in economics and finance, vol 563. Stockholm School of Economics, Stockholm
Mandelbrot B (1963) The variation of certain speculative prices. J Bus 36:394–419
Markowitz H (1959) Portfolio selection: efficient diversification of investments. Wiley, New York
Meitz M, Saikkonen P (2012) Parameter estimation in nonlinear AR.GARCH models. Econ Theory 27(6):1236–1278
Merton RC (1973) An intertemporal capital asset pricing model. Econometrica 41:867–888
Mikosch T, Rezapur M (2012) Stochastic volatility models with possible extremal clustering. Bernoulli (to appear)
Nelson DB (1990a) Stationary and persistence in the GARCH(1,1) model. Econ Theory 6:318–334
Nelson DB (1990b) ARCH models as diffusion approximations. J Econ 45:7–38
Nelson DB (1991) Conditional heteroskedasticity in asset returns: a new approach. Econometrica 59:347–370
Nelson DB, Cao CQ (1992) Inequality constraints in the univariate GARCH model. J Bus Econ Stat 10:229–235
Newey W, Steigerwald D (1997) Asymptotic bias for quasi-maximum-likelihood estimators in conditional heteroscedasticity models. Econometrica 65:587–599
Nocedal J, Wright SJ (1999) Numerical optimization. Springer, New York
Pantula SG (1986) Modeling the persistence of conditional variances: a comment. Econ Rev 5:71–74
Poon SH (2005) A practical guide to forecasting financial market volatility. Wiley, New York
Rodriguez MJ, Ruiz M (2012) Revisiting several popular GARCH models with leverage effect: differences and similarities. J Financ Econ 10(4):637–668
Rogers LCG, Satchell SE (1991) Estimating variance from high, low and closing prices. Ann Appl Probab 1(4):504–512
Sentana E (1995) Quadratic ARCH models. Rev Econ Stud 62:639–661
Straumann D (2005) Estimation in conditionally heteroscedastic time series models. Lecture notes in statistics, vol 181. Springer, Berlin
Stroock DW, Varadhan SRS (1979) Multidimensional diffusion processes. Springer, New York
Taylor SJ (1986) Modelling financial time series. Wiley, New York
Taylor SJ (2005) Asset price dynamics, volatility, and prediction. Princeton University Press, Princeton
Terasvirta T (2009) An introduction to univariate GARCH models. In: Andersen TG, Davis RA, Kreiss JP, Mikosch T (eds) Handbook of financial time series. Springer, New York, pp 17–42
Terasvirta T, Zhao Z (2011) Stylized facts of return series, robust estimates and three popular models of volatility. Appl Financ Econ 21(1–2):67–94
Venkataraman S (1997) Value at risk for a mixture of normal distributions: the use of quasi-Bayesian estimation techniques. Econ Perspect Federal Reserve Bank of Chicago 21:2–13
Wang Y (2002) Asymptotic nonequivalence of GARCH models and diffusions. Ann Stat 30:754–783

Wintenberger O (2013) Continuous invertibility and stable QML estimation of EGARCH(1,1) model. Scand J Stat 40(4): 846–867
Wold H (1938) A study in the analysis of stationary time series. Almqvist and Wiksell, Uppsala
Yang M (2011) Volatility feedback and risk premium in GARCH models with generalized hyperbolic distributions. Stud Nonlinear Dyn Econ 15(3):1–21
Zakoian JM (1994) Threshold heteroskedastic models. J Econ Dyn Control 18:931–955

From Time Series of Returns to Option Prices: The Stochastic Discount Factor Approach

3

In the perfect and unrealistic Black and Scholes (1973) world, the dynamics $(S_t)_{t\in[0,T]}$ of the risky asset, under the historical probability $\mathbb{P}$, is given by the following stochastic differential equation:

$$dS_t = \mu S_t dt + \sigma S_t dW_t \tag{3.1}$$

where $(W_t)_{t\in[0,T]}$ is a standard Brownian motion under $\mathbb{P}$. In this case, there is no ambiguity in the definition the arbitrage-free price of any European contingent claim with maturity T. In fact, in this complete market which is set in continuous time, this value is none other than the value of any replicating portfolio. Moreover, prices may be expressed in terms of conditional expectations under a unique equivalent martingale measure Q whose density with respect to the historical probability is given by the Girsanov theorem

$$\frac{dQ}{d\mathbb{P}} = e^{-\frac{\mu-r}{\sigma}W_T - \left(\frac{\mu-r}{\sigma}\right)^2 \frac{T}{2}} \tag{3.2}$$

where r is the constant and continuously compound risk-free rate. Unfortunately, as we have seen in Sect. 2.1, the restrictive underlying hypotheses (constant volatility, independent increments, Gaussian log-returns, etc...) are questioned by many empirical studies and GARCH models appear as excellent alternative solutions to potentially overcome some well-documented systematic biases associated with the Black and Scholes model.

Nevertheless, conditional distributions of returns and volatility specifications are not the only issues one should pay attention to when pricing contingent claims. In fact, modeling asset prices using realistic discrete time volatility structures and continuous distributions gives rise to incompleteness: hedging arguments become impossible in this infinite state space setup and there are contingent claims that cannot be replicated using a self-financing portfolio made only out of a bond and the risky asset. There exist, in general, more than one pricing measure compatible

C. Chorro et al., *A Time Series Approach to Option Pricing*,
DOI 10.1007/978-3-662-45037-6_3

with the no-arbitrage assumption. A great challenge is to be able to select tractable candidates for their strong economic foundations and empirical performances. In the case of GARCH models with Gaussian innovations, a major step was done in Duan (1995) where an extension of the so-called Risk Neutral Valuation Relationship (RNVR) pioneered by Rubinstein (1976) and developed by Brennan (1979) was proposed to give a strong economical foundation to the Gaussian GARCH option pricing model independently of the choice of the GARCH volatility structure. As remarked by Heston and Nandi (2000), this economic principle is equivalent to consider that the classical Black and Scholes (1973) formula holds for call options with one period to expiration.[1] This framework presents the concrete possibility of correcting some of the price biases associated with the Black and Scholes model.[2] Nevertheless, as observed in Sect. 2.2.5, GARCH models with Gaussian innovations have shown their inability to take into account all the mass in the tails and the asymmetry that characterize the distributions of daily log-returns. In particular, in these models, single-period innovations are Gaussian partly explaining why they usually fail to reproduce the short term behavior of equity option smiles. During the last decade, researchers have intensively investigated the way to extend the Duan's option pricing model to incorporate in GARCH residuals the skewness and leptokurtosis observed in financial datasets.

The aim of this chapter is to present intuitively and theoretically these recent advances and to compare the classical approaches for the pricing of European contingent claims in the GARCH framework especially when the innovations are not Gaussian. In particular, we derive all the risk-neutral dynamics implemented in Chap. 4. The technical proofs are postponed to the end of the chapter for reader's convenience.

3.1 Description of the Economy Under the Historical Probability

We consider in this chapter a discrete time economy with a time horizon $T \in \mathbb{N}^*$ consisting of a risk-free zero-coupon bond and a stock (the risky asset). Let $(z_t)_{t\in\{1,\dots,T\}}$ be independent and identically distributed random variables defined

[1] In their paper (see also Sect. 2.7), Heston and Nandi also provided a particular GARCH structure that may be seen as a discrete time counterpart of the so-called Heston (1993) stochastic volatility model replicating one of key features of this continuous time model: the fact that the characteristic function of the log-returns under the risk-neutral distribution had a closed-form expression (see Sect. 3.7). This result is particularly interesting because it underlines that GARCH option pricing models may also be seen as competitive discrete time approximations of the classical models used in continuous time finance with the great advantage that they are easy to estimate because the resulting filtering problem is simple. In particular, they may represent interesting and efficient alternative options to Euler approximation schemes (see Duan et al. 2006 and Lindner 2009).

[2] For example, it is now well-documented that the Black and Scholes model underprices out of the money and short maturity options (see Black 1975) and is not able to reproduce the so-called U-shaped relationship between implied volatility and strike price as remarked in Rubinstein (1985).

on a complete probability space $(\Omega, \mathcal{A}, \mathbb{P})$. We define the associated information filtration by $(\mathcal{F}_t)_{t\in\{0,\dots,T\}}$ where $\mathcal{F}_0 = \{\emptyset, \Omega\}$ and $(\mathcal{F}_t = \sigma(z_u; 1 \le u \le t))_{t\in\{1,\dots,T\}}$. We assume that under the historical probability $\mathbb{P}$, the dynamics of the bond price process $(B_t)_{t\in\{0,\dots,T\}}$ and of the stock price process $(S_t)_{t\in\{0,\dots,T\}}$ are given by

$$B_t = B_{t-1}e^r, \quad B_0 = 1, \tag{3.3}$$

where r is the corresponding risk-free rate expressed on a daily basis, supposed to be constant and

$$Y_t = log\left(\frac{S_t}{S_{t-1}}\right) = r + m_t + \underbrace{\sqrt{h_t}z_t}_{\varepsilon_t}, \quad S_0 = s \in \mathbb{R}_+, \tag{3.4}$$

where z_t follows an arbitrary distribution $D(0,1)$ with mean 0 and variance 1. In Eq. (3.4), we consider a general time varying excess of return m_t that is $\mathcal{F}_{t-1}$ measurable and implicitly depends on the constant unit risk premium λ_0 (see Sect. 2.5). For the conditional variance of the log-returns we choose

$$h_t = F(z_{t-1}, h_{t-1}) \tag{3.5}$$

where $F : \mathbb{R}^2 \to \mathbb{R}_+$ has a general form that is not specified for the moment but is compatible with realistic GARCH(1,1)-type volatility models introduced in Chap. 2. From now on, we suppose that h_0 is a constant.[3]

Remark 3.1.1 Taking $F \equiv \sigma^2$, $m_t = \lambda_0\sqrt{h_t} - \frac{1}{2}h_t$ with $\lambda_0 = \frac{\mu-r}{\sigma}$ we recover the discrete time counterpart of the Black and Scholes economy (3.1) once z_t follows a $\mathcal{N}(0,1)$.

This general specification allows in particular the log-returns to have fat-tailed and skewed conditional distributions. This feature is important because, as remarked in Chap. 2, empirical studies have proved that financial data continue to exhibit fat tails and skewness even if a GARCH filter is applied (see Fig. 2.5). However, having option pricing in mind, the flexibility to use continuous distributions (and so to work in an infinite state space setup) in discrete time dynamics gives typically rise to incompleteness (see Jacod and Shiryaev 1998). Classical hedging arguments become invalid and it is not possible to define prices for contingent claims creating riskless portfolio. In the next section, we briefly remind the notion of arbitrage-free prices in the discrete time setting and its link with the existence of so-called stochastic discount factors.

[3] In this case, the information filtration $(\mathcal{F}_t)_{t\in\{1,\dots,T\}}$ is also generated by the log-returns $(Y_t)_{t\in\{1,\dots,T\}}$ and the weak and strong market efficiency hypotheses coincide (see Elliott and Madan 1998).

3.2 Option Pricing in Discrete Time

We now turn our attention on the pricing of a European contingent claim with maturity T that is, in our framework, perfectly characterized by a random variable $C_T \in L^1(\Omega, \mathcal{F}_T, \mathbb{P})$, called the payoff function, that represents the value of this financial contract at time T.

Example 3.2.1 When $C_T = Max(S_T - K, 0)$ (resp. $C_T = Max(K - S_T, 0)$), the associated European contingent claim is none other than a classical European call (resp. put) option that gives its holder the right, and not the obligation, to buy (resp. sell) at time T one risky asset for the strike price K fixed at time $t = 0$.

When $(X_t)_{t\in\{0,\dots,T\}}$ is a discrete time stochastic process, we will denote by $(\tilde{X}_t)_{t\in\{0,\dots,T\}}$ the associated discounted process defined by $\tilde{X}_t = \frac{X_t}{B_t}$.

Hypothesis: In this section we suppose the existence of an equivalent martingale measure (EMM) Q: Q is a probability defined on the space $(\Omega, \mathcal{F}_T)$, equivalent to $\mathbb{P}$ and such that $(\tilde{S}_t)_{t\in\{0,\dots,T\}}$ is, under Q, a martingale with respect to the information filtration $(\mathcal{F}_t)_{t\in\{0,\dots,T\}}$. We denote by $\frac{dQ}{d\mathbb{P}}$ its Radon–Nikodym derivative. In particular we have $\forall t \in \{0, \dots, T-1\}$

$$E_Q[S_{t+1} \mid \mathcal{F}_t] = e^r S_t. \tag{3.6}$$

Remark 3.2.1 The links between the existence of an EMM and the absence of arbitrage opportunities (AAO) in financial models is a classical topic for frictionless markets (see Delbaen and Schachermayer 2006 for an overview). Contrary to the continuous time case, for finitely many trading dates and assets the equivalence between these two notions (known as the "Fundamental Theorem of Asset Pricing") was proved, without any technical restriction, in Dalang et al. (1990) for general Ω and in Harrison and Pliska (1981) for finite Ω with profound different approaches (we refer the reader to Back and Pliska (1991) and Schachermayer (1992) for counterexamples in the implication AAO $\Rightarrow$ EMM for infinite trading dates or assets). We will see from Sects. 3.3 to 3.6 how to practically select such an EMM in our framework.

3.2.1 Arbitrage-Free Price of a European Contingent Claim

We present in this section the definition of an arbitrage-free price as introduced in Kreps (1981). It requires to remind some notions that are now well known in the financial literature, then, we omit the classical proofs and refer the reader to Delbaen and Schachermayer (2006) for more details.

Once a European contingent claim with payoff C_T is fixed, we consider an extended financial market that includes a bond and a risky asset, supposing that it is possible to buy or to sell at any time any quantity of the European contingent claim. We denote by $(C_t)_{t\in\{0,\ldots,T\}}$ the adapted price process (to determine) of this contingent claim and we suppose that this new market is without frictions meaning that:

(i) There are no transaction costs or taxes.
(ii) It is possible to borrow and lend the bond and to short sell the other assets.
(iii) All securities are perfectly divisible.
(iv) Assets do not pay dividends.

Now we introduce the notion of financial strategies that is compatible with the four preceding points.

Definition 3.2.1 (a) A financial strategy is defined as a $\mathbb{R}^3$ valued and adapted stochastic process $(H_t = (\theta_t^0, \theta_t, \theta_t^C))_{t\in\{0,\ldots,T\}}$. We associate to any financial strategy a financial portfolio containing between t and $t+1$ the quantity $(\theta_t^0, \theta_t, \theta_t^C)$ of bond, risky asset and contingent claim. We denote by V_t the value at time $t \in \{0,\ldots,T\}$ of such a portfolio given by

$$V_t = \theta_t^0 B_t + \theta_t S_t + \theta_t^C C_t. \tag{3.7}$$

(b) A financial strategy is self-financing if the consumption process $(\overline{c}_t)_{t\in\{1,\ldots,T\}}$ is equal to zero: $\forall t \in \{1,\ldots,T\}$,

$$-\overline{c}_t = V_t - \left(\theta_{t-1}^0 B_t + \theta_{t-1} S_t + \theta_{t-1}^C C_t\right) = 0. \tag{3.8}$$

(c) A self-financing financial strategy is an arbitrage opportunity (AO) if

$$V_0 = 0,\ \mathbb{P}(V_T \geq 0) = 1 \textit{ and } \mathbb{P}(V_T > 0) > 0. \tag{3.9}$$

In other terms, starting from zero (no initial investment), we never lose money at time T and there are real opportunities of profit.

Remark 3.2.2 (a) The measurability property fulfilled by the process $(H_t)_{t\in\{0,\ldots,T\}}$ is intuitive: a financial decision (sell or buy the assets) taken at time t is naturally based on the financial information available at this time i.e. $\mathcal{F}_t$. The fact that $(H_t)_{t\in\{0,\ldots,T\}}$ have values in $\mathbb{R}^3$ is a consequence of (ii) (negative values are allowed) and (iii) (non integer values are allowed). Equation (3.7), expressing that values are none other than quantities multiplied by prices, is a consequence of (i) and (iv).

(b) It is straightforward (see Delbaen and Schachermayer 2006) that Eq. (3.8) is equivalent to

$$\tilde{V}_t = V_0 + \sum_{k=0}^{t-1} \theta_k (\tilde{S}_{k+1} - \tilde{S}_k) + \sum_{k=0}^{t-1} \theta_k^C (\tilde{C}_{k+1} - \tilde{C}_k),\ \forall t \in \{1,\ldots,T\}. \tag{3.10}$$

Thus, the self-financing condition implies that changes in the value of the portfolio only come from changes in the assets values. Between 0 and T, we do not have the right to consume or to invest, the only possibility is to rebalance the portfolio respecting the budget constraints.

(c) The concept of AO depends on the choice of the historical probability $\mathbb{P}$. Nevertheless, if Q is a probability measure equivalent to $\mathbb{P}$, we may replace $\mathbb{P}$ by Q in the preceding definition.

Now we are able to define the notion of arbitrage-free price.

Definition 3.2.2 The process $(C_t)_{t\in\{0,\dots,T\}}$ is an arbitrage-free price process if there are no AO in the extended market.

From Eq. (3.10), we deduce that $(C_t)_{t\in\{0,\dots,T\}}$ is an arbitrage-free price process if $(\tilde{C}_t)_{t\in\{0,\dots,T\}}$ is a martingale under an EMM Q. In fact, in this case, $(\tilde{V}_t)_{t\in\{0,\dots,T\}}$ is also a martingale under Q and the conditions

$$V_0 = 0,\ \mathbb{P}(V_T \geq 0) = 1 \text{ and } \mathbb{P}(V_T > 0) > 0$$

are incompatible because $E_Q[\tilde{V_T}] = V_0$. In particular, we have, as an immediate consequence, the following proposition:

Proposition 3.2.1 *When Q is an EMM, the process $(C_t)_{t\in\{0,\dots,T\}}$ defined by*

$$C_t = E_Q\left[\frac{B_t C_T}{B_T} \mid \mathcal{F}_t\right] \tag{3.11}$$

is an arbitrage-free price process.

In the preceding proposition, the arbitrage-free price process depends on the choice of the EMM Q because, in our incomplete setting, European contingent claims are not in general replicable (see Delbaen and Schachermayer 2006). Moreover (see Proposition A.4 of the Appendix), it is possible to express C_t using a conditional expectation under the historical probability $\mathbb{P}$, in fact,

$$C_t = E_{\mathbb{P}}\left[\frac{B_t C_T}{B_T}\frac{L_T}{L_t} \mid \mathcal{F}_t\right] \tag{3.12}$$

where

$$L_t = E_{\mathbb{P}}\left[\frac{dQ}{d\mathbb{P}} \mid \mathcal{F}_t\right] \tag{3.13}$$

is a martingale under $\mathbb{P}$.

3.2.2 The Stochastic Discount Factor

Following Harrison and Kreps (1979) and Hansen and Richard (1987), we introduce in this section the definition of the stochastic discount factor (SDF).

Definition 3.2.3 Let $(M_t)_{t\in\{1,\ldots,T\}}$ be a positive process adapted to the information filtration $(\mathcal{F}_t)_{t\in\{1,\ldots,T\}}$. This process is called a one-period stochastic discount factor process if the following relations hold $\forall t \in \{0,\ldots,T-1\}$:

$$E_{\mathbb{P}}\left[\frac{B_{t+1}M_{t+1}}{B_t} \mid \mathcal{F}_t\right] = 1 \tag{3.14}$$

$$E_{\mathbb{P}}\left[\frac{S_{t+1}}{S_t}M_{t+1} \mid \mathcal{F}_t\right] = 1. \tag{3.15}$$

From the law of iterated expectations (see Proposition A.2 of the Appendix), it is easy to see that

$$E_{\mathbb{P}}\left[\frac{S_T}{S_t}M_{t,T} \mid \mathcal{F}_t\right] = 1 \tag{3.16}$$

where

$$M_{t,T} = \prod_{k=t+1}^{T} M_k \tag{3.17}$$

Let us mention that in the literature the process $(e^r M_t)_{t\in\{1,\ldots,T\}}$ is also known as a one-period pricing kernel process.

When comparing Eqs. (3.14)–(3.15) and (3.12)–(3.13) we can see clearly the link between the notion of EMM and the preceding definition, through the relations

$$e^{-r}\frac{L_{t+1}}{L_t} = M_{t+1} \tag{3.18}$$

$$\frac{dQ}{d\mathbb{P}} = e^{rT}\prod_{k=1}^{T} M_k. \tag{3.19}$$

Thus, the construction of an EMM (and thus, the obtaining of arbitrage-free prices) is equivalent to the specification of a one-period SDF process.

Remark 3.2.3 In the classical Black and Scholes economy it is easy to see that the unique EMM coming from the Girsanov theorem (see Eq. (3.2)) is associated with the unique one-period SDF process

$$M_{t+1} = e^{-r-(\frac{\mu-r}{\sigma^2})log(\frac{S_{t+1}}{S_t})+(\frac{\mu-r}{2\sigma^2})(\mu+r-\sigma^2)} \tag{3.20}$$

that is an exponential affine function of the log-returns. In this Markovian setting, M_{t+1} is none other than the ratio $e^{-r}\frac{q_t(Y_{t+1})}{p_t(Y_{t+1})}$ where p_t (resp. q_t) is the density of $Y_{t+1} = log(\frac{S_{t+1}}{S_t})$ under $\mathbb{P}$ (resp. Q). More generally, let us consider a family $(c_t)_{t\in\{1,\dots,T\}}$ of state variables fulfilling $S_t = h(c_t, S_{t-1})$ and such that $\mathcal{F}_t = \sigma(c_1,\dots,c_t)$.[4] We deduce from

$$E_Q[S_{t+1} \mid \mathcal{F}_t] = \int_{\mathbb{R}} h(y, S_t)\frac{q(y \mid c_1,\dots,c_t)}{p(y \mid c_1,\dots,c_t)}p(y \mid c_1,\dots,c_t)dy$$

a natural candidate for the one period SDF process:

$$e^{-r}\frac{q(c_{t+1} \mid c_1,\dots,c_t)}{p(c_{t+1} \mid c_1,\dots,c_t)} = M_{t+1} \tag{3.21}$$

where $p(y \mid c_1,\dots,c_t)$ (resp. $q(y \mid c_1,\dots,c_t)$) is the conditional density of c_{t+1} given $\mathcal{F}_t$ under $\mathbb{P}$ (resp. under Q). Similarly,

$$M_{t,T} = e^{-r(T-t)}\prod_{k=t}^{T-1}\frac{q(c_{k+1} \mid c_1,\dots,c_k)}{p(c_{k+1} \mid c_1,\dots,c_k)} = e^{-r(T-t)}\frac{q(c_{t+1},\dots,c_T \mid c_1,\dots,c_t)}{p(c_{t+1},\dots,c_T \mid c_1,\dots,c_t)}$$

where $p(y_{t+1},\dots,y_T \mid c_1,\dots,c_t)$ (resp. $q(y_{t+1},\dots,y_T \mid c_1,\dots,c_t)$) is the conditional density of $(c_{t+1},\dots,c_T)$ given $\mathcal{F}_t$ under $\mathbb{P}$ (resp. under Q).

From the seminal paper of Rubinstein (1976), it is classical that relation (3.20) may be justified by an equilibrium-based asset pricing model under power utility assumptions for preferences. In the next section, following Cochrane (2001) we remind how to obtain a SDF from economic principles and present the Duan (1995) "Local Risk Neutral Valuation Relationship" (LRNVR) for GARCH models with Gaussian innovations.

[4]In the case of the discrete time economy introduced in Sect. 3.1, we can take $c_t = Y_t$ and $h(c,s) = se^c$.

3.2.3 Economic Interpretation: The CCAPM Model

In this section, we consider a Lucas (1978) economy with a risk averse representative agent. His preferences are modeled by a strictly increasing, time separable and additive one-period utility function $u(c_t)$, where c_t represents the aggregate consumption at time $t \in \{0, \dots, T-1\}$. Between t and $t+1$, this investor can freely buy or sell as much of the risky asset and of the bond as he wishes. If we denote by e_t its original endowment at time t, ξ (resp. $\tilde{\xi}$) the quantity of risky asset (resp. bond) he decides to buy and ρ the intertemporal psychological discount rate which measures the degree of preference of the agent to consume now rather than a period later, the standard utility maximization problem may be written as

$$\underset{\xi,\tilde{\xi}}{Max}\ u(c_t) + e^{-\rho} E_{\mathbb{P}}[u(c_{t+1}) \mid \mathcal{F}_t] \tag{3.22}$$

under the budget constraint at time t

$$e_t = c_t + \xi S_t + \tilde{\xi} B_t \tag{3.23}$$

and the Walrasian property (everything has to be consumed at time $t+1$)

$$c_{t+1} = e_{t+1} + \xi S_{t+1} + \tilde{\xi} B_{t+1}. \tag{3.24}$$

Partial derivatives with respect to ξ and $\tilde{\xi}$ being equal to zero, we obtain the so-called Euler conditions:

$$e^{-r} = e^{-\rho} E_{\mathbb{P}}\left[\frac{u'(c_{t+1})}{u'(c_t)} \mid \mathcal{F}_t\right] \tag{3.25}$$

$$S_t = e^{-\rho} E_{\mathbb{P}}\left[\frac{u'(c_{t+1})}{u'(c_t)} S_{t+1} \mid \mathcal{F}_t\right]. \tag{3.26}$$

Thus, from Definition 3.2.3, we identify the equilibrium one-period SDF:

$$M_{t+1} = e^{-\rho}\frac{u'(c_{t+1})}{u'(c_t)}. \tag{3.27}$$

Remark 3.2.4 In the case of an expected utility maximizer agent, Arrow and Pratt (see for instance Pratt 1964) introduced the so-called absolute risk aversion coefficient $\rho_t(c_{t+1})$ between t and $t+1$ defined by

$$\rho_t(c_{t+1}) = -\frac{u''(c_{t+1})}{u'(c_{t+1})}.$$

This definition may be extended to any SDF remarking that Eq. (3.27) implies that

$$-\frac{u''(c_{t+1})}{u'(c_{t+1})} = -\frac{\partial log M_{t+1}}{\partial c_{t+1}}.$$

In particular using Eq. (3.21) we find

$$\rho_t(c_{t+1}) = \frac{q'(c_{t+1} \mid c_1, \ldots, c_t)}{q(c_{t+1} \mid c_1, \ldots, c_t)} - \frac{p'(c_{t+1} \mid c_1, \ldots, c_t)}{p(c_{t+1} \mid c_1, \ldots, c_t)}.$$

Thus, the preceding relation gives a general framework to infer the Arrow-Pratt absolute risk aversion coefficient from vanilla option prices and historical dynamics (see for example Aît-Sahalia and Lo 1998, 2000; Jackwerth 2000; Rosenberg and Engle 2002 and Badescu et al. 2008).

Now, let us assume that the dynamics of the risky asset in the Lucas economy is given by Eq. (3.4), where the innovations z_t are Gaussian, and that the underlying utility function is isoelastic[5]:

$$u(x) = \frac{x^{1-R} - 1}{1 - R} \tag{3.28}$$

where $R > 1$ is called the relative risk aversion coefficient. Moreover, we suppose that, conditionally to $\mathcal{F}_{t-1}$,

$$(Y_t, \Delta log(c_t) = log(c_t) - log(c_{t-1}))$$

is a Gaussian vector. In other words, we can equivalently write that

$$\Delta log(c_t) = A_t^0 + A_t^1 Y_t + A_t^2 \eta_t \tag{3.29}$$

where $(A_t^0)_{t\in\{1,\ldots,T\}}$, $(A_t^1)_{t\in\{1,\ldots,T\}}$ and $(A_t^2)_{t\in\{1,\ldots,T\}}$ are three predictable processes (see Definition A.2) and where $(\eta_t)_{t\in\{1,\ldots,T\}}$ is a sequence of $i.i.d$ $\mathcal{N}(0, 1)$ such that η_t is independent of z_t given $\mathcal{F}_{t-1}$.[6] In this setting, we deduce from Eq. (3.27) that

$$M_t = e^{-\rho - R\Delta log(c_t)} = e^{-\rho - R(A_t^0 + A_t^1 Y_t + A_t^2 \eta_t)} \tag{3.30}$$

[5]It means that, in this case, the relative risk aversion coefficient $c_{t+1}\rho_t(c_{t+1})$ is a constant.

[6]This may be seen as a conditional version of the following elementary result concerning two-dimensional Gaussian vectors: When (X, Y) is a Gaussian vector, $\exists(a, b) \in \mathbb{R}^2$ such that $X - a - bY$ is a centered Gaussian random variable independent of Y. The proof derives easily from the fact that the orthogonal projection $\Pi(X)$ of X on the vector space generated by the random variables 1 and Y is an affine function of Y and that $X - \Pi(X)$ is, by construction, a Gaussian random variable independent of Y.

and from Eq. (3.19) that

$$\frac{dQ}{d\mathbb{P}} = e^{(r-\rho)T} \prod_{k=1}^{T} e^{-R(A_k^0 + A_k^1 Y_k + A_k^2 \eta_k)}. \tag{3.31}$$

To find the dynamics of the risky asset under this new probability we compute the conditional moment generating function of the log-returns. Thus, we obtain from Proposition A.4 of the Appendix that

$$\mathbb{G}^{Q}_{Y_t|\mathcal{F}_{t-1}}(u) = E_Q[e^{uY_t} \mid \mathcal{F}_{t-1}] = E_{\mathbb{P}}\left[e^{uY_t + (r-\rho) - R(A_t^0 + A_t^1 Y_t + A_t^2 \eta_t)} \mid \mathcal{F}_{t-1}\right]. \tag{3.32}$$

Using (3.25) and (3.26) we have

$$\mathbb{G}^{Q}_{Y_t|\mathcal{F}_{t-1}}(u) = \frac{\mathbb{G}^{\mathbb{P}}_{Y_t|\mathcal{F}_{t-1}}(u - RA_t^1)}{\mathbb{G}^{\mathbb{P}}_{Y_t|\mathcal{F}_{t-1}}(-RA_t^1)} \tag{3.33}$$

where

$$\frac{G^{\mathbb{P}}_{Y_t|\mathcal{F}_{t-1}}(1 - RA_t^1)}{G^{\mathbb{P}}_{Y_t|\mathcal{F}_{t-1}}(-RA_t^1)} = e^r.$$

The innovations being Gaussian under $\mathbb{P}$,

$$G^{\mathbb{P}}_{Y_t|\mathcal{F}_{t-1}}(u) = e^{u(r+m_t) + \frac{u^2 h_t}{2}}$$

thus,

$$RA_t^1 = \frac{m_t}{h_t} + \frac{1}{2}$$

and

$$\mathbb{G}^{Q}_{Y_t|\mathcal{F}_{t-1}}(u) = e^{u(r - \frac{h_t}{2}) + \frac{u^2 h_t}{2}}. \tag{3.34}$$

In particular, the conditional distribution of Y_t given $\mathcal{F}_{t-1}$ under the equilibrium EMM Q is a $\mathcal{N}(r - \frac{h_t}{2}, h_t)$. As we are going to see below, this risk-neutral dynamics is related to the Duan's (1995) approach of option pricing in a Gaussian GARCH setting.

In a seminal paper, Duan (1995) proposes the following extension of the risk-neutral valuation relationship pioneered by Rubinstein (1976) and developed by Brennan (1979) and gives a strong economical foundation to the GARCH option pricing model with Gaussian innovations. Moreover, the dynamics under the pricing measure is explicitly described.

Definition 3.2.4 (Duan 1995) An EMM Q fulfills the "Locally Risk Neutral Valuation Relationship" (LRNVR) if

(1) Given $\mathcal{F}_{t-1}$, Y_t follows a Gaussian distribution under Q,
(2) $Var_Q[Y_t \mid \mathcal{F}_{t-1}] = Var_{\mathbb{P}}[Y_t \mid \mathcal{F}_{t-1}]$.

Proposition 3.2.2 (Duan 1995) *If Q fulfills the LRNVR, the dynamics of the log-returns under Q is given by*

$$Y_t = r - \frac{h_t}{2} + \sqrt{h_t}\xi_t, \quad S_0 = s, \tag{3.35}$$

where

$$h_t = F\left(\xi_{t-1} - \frac{m_{t-1}}{\sqrt{h_{t-1}}} - \frac{\sqrt{h_{t-1}}}{2}, h_{t-1}\right) \tag{3.36}$$

and where the ξ_t are i.i.d $\mathcal{N}(0,1)$ under Q.

According to Eq. (3.34), we see that the equilibrium EMM given by (3.31) fulfills the LRNVR and so the associated risk-neutral dynamics is given by Proposition 3.2.2.

We end this section reminding some other economic conditions for the LRNVR to hold. The proof is omitted because it follows exactly the same lines as the isoelastic utility case.

Proposition 3.2.3 (Duan 1995) *In the preceding Lucas type economy, the equilibrium one-period SDF process is associated to an EMM that fulfills the LRNVR under any of the three following conditions:*

(1) u is linear.
(2) The utility function has the form (3.28) and, given $\mathcal{F}_{t-1}$, $(Y_t, \Delta log(c_t))$ is a Gaussian vector.
(3) The utility function has an exponential form $u(x) = \frac{1-e^{-Rx}}{R}$, where $R > 0$ is the coefficient of absolute risk aversion, and, given $\mathcal{F}_{t-1}$, $(Y_t, \Delta c_t = c_t - c_{t-1})$ is a Gaussian vector.

It is important to remark that, in Proposition 3.2.2, the conditional mean m_t and the GARCH structure F can actually be specified in a fairly general way in (3.4) and (3.5). They may be chosen for technical reasons (for example to obtain a closed-form solution for vanilla option prices to calibrate the model via option datasets as in Heston and Nandi 2000) or to cope with classical stylized facts. In particular, in Duan's (1995) paper, a GARCH(1,1) volatility structure is used

$$\begin{cases} m_t = \lambda_0\sqrt{h_t} - \frac{h_t}{2} \\ F(z_{t-1}, h_{t-1}) = a_0 + a_1(\varepsilon_{t-1})^2 + b_1 h_{t-1} \end{cases} \tag{3.37}$$

while in the affine Heston and Nandi (2000) model a leverage effect parameter is introduced under $\mathbb{P}$

$$\begin{cases} m_t = \lambda_0 h_t \\ F(z_{t-1}, h_{t-1}) = a_0 + a_1(z_{t-1} - \gamma\sqrt{h_{t-1}})^2 + b_1 h_{t-1}. \end{cases} \tag{3.38}$$

In both cases, we see from Proposition 3.2.2 that the dynamics under Q are compatible with the stylized fact that the risk-neutral distribution of returns is more negatively skewed than the physical distribution. Nevertheless, while the leverage parameters of the preceding models create negative skewness in large horizon, single-period innovations are symmetric. In particular, they usually fail to capture the short term behavior of equity option smiles. As discussed in Chap. 2, the importance of asymmetry and heavy tails inside financial time series is now well-documented in the literature, as in Bouchaud and Potters (2003) or Embretchs et al. (2005), especially for pricing issues. In the next section, we try to relax the Gaussian hypothesis for the innovations of the risky asset introducing more flexible specifications of the SDF.

We have seen that each econometric asset pricing model is based on three key ingredients:

(i) the historical dynamics of the underlying asset,
(ii) the SDF that takes into account risk corrections,
(iii) the risk-neutral dynamics used to price contingent claims.

The three elements cannot be defined independently and following Bertholon et al. (2008) three strategies are available:

1. The direct approach: start from the historical distribution, assume a shape for the risk premium and the risk-neutral distribution is a by-product of the approach.
2. The risk-neutral constrained direct modelling: choose a specification for the risk-neutral and the historical distribution and obtain the risk premium as a by-product.
3. The back modelling approach: choose a specification for the risk-neutral distribution and the risk aversion and obtain the historical distribution as a by-product.

From a statistical point of view, the approach (1) is the most relevant in particular for illiquid markets where risk-neutral dynamics are difficult to calibrate. It will be adopted in the following: Fitting a relevant econometric model for the log-returns and specifying a particular form for the SDF we are going to recover the implied risk-neutral dynamics. Nevertheless, as we have seen in Remark 3.2.4, in liquid markets the approach (2) may be relevant to extract risk aversion from option prices. For example, in Barone-Adesi et al. (2008) the authors calibrate independently two non-parametric GARCH processes on stock and option prices and obtain a SDF compatible with economic theory.

We present in the next sections the classical specifications of the SDF.

3.3 The Extended Girsanov Principle

3.3.1 Definition and Properties

In the classical Black and Scholes (1973) model, the unique EMM is obtained from the Girsanov theorem shifting the initial distribution of the log-returns (it is also-called in the literature the mean correction martingale measure). The LRNVR of Duan (1995) is essentially based on the same underlying idea. In this section, we present a general method to obtain for discrete time models a SDF compatible with shift methods even if the innovations are not Gaussian.

First of all, let us state a simple probabilistic result that is the key stone of this approach. It may be seen intuitively as a static version of the Girsanov theorem:

Lemma 3.3.1 *Let e^Z be a real random variable defined on the probability space $(\Omega, \mathcal{A}, \mathbb{P})$ that has a strictly positive density d with respect to the Lebesgue measure. For all $m \in \mathbb{R}_+$, there exists a unique probability measure $\mathbb{Q}_m$ defined on $(\Omega, \sigma(e^Z))$*[7] *and equivalent to $\mathbb{P}$ restricted to $\sigma(e^Z)$ such that the distribution of e^Z under $\mathbb{Q}_m$ is equal to the distribution of e^{Z-m} under $\mathbb{P}$. Moreover, we have*

$$\frac{d\mathbb{Q}_m}{d\mathbb{P}} = \frac{d(e^{Z+m})}{d(e^Z)e^{-m}} = \frac{\tilde{d}(e^Z)e^m}{\tilde{d}(e^{Z-m})} \tag{3.39}$$

where $\tilde{d}$ is the density of e^Z under $\mathbb{Q}_m$. In particular, when $m = log(E_{\mathbb{P}}[e^Z])$ we have $E_{\mathbb{Q}_m}[e^Z] = 1$.

Remarking that, for discrete time models with infinite state space, it is not possible to use some classical tools of continuous time finance as the minimal martingale measure of Föllmer and Schweiser (1991),[8] Elliott and Madan (1998) proposed a very simple way to choose an EMM based on a Girsanov-type transformation that preserves the distribution of returns after the change of measure by only shifting

[7] As introduced in Remark A.1 of the Appendix, $\sigma(e^Z)$ is the smallest σ-algebra on Ω that makes the random variable e^Z measurable. In particular, a real random variable X defined on Ω is $\sigma(e^Z)$ measurable if and only if it is of the form $X = h(e^Z)$ where $h : \mathbb{R} \to \mathbb{R}$ is Borelian.

[8] In Ortega (2012) it is proved that the range of GARCH situations in which the minimal martingale measure exists is rather limited.

the conditional mean. It is based on a multiplicative Doob decomposition of the discounted stock price into a predictable finite variation component and a martingale one. Under mild integrability conditions, Elliott and Madan built a risk-neutral probability under which the conditional distribution of the discounted stock price is equal to the conditional distribution of its martingale component under the historical probability. This discrete time method is also well adapted to exponential Lévy models (see Schoutens 2003). Now, we recall briefly the main lines of this approach.

We denote by $(\mathcal{G}_t = \sigma(Y_u; 1 \leq u \leq t))_{t \in \{0,1,\dots,T\}}$ the filtration generated by the log-returns. In our GARCH framework given by Eqs. (3.4) and (3.5) the preceding filtration is none other than $(\mathcal{F}_t)_{t \in \{0,1,\dots,T\}}$ once h_0 is known. Now we decompose the discounted stock price process $(\tilde{S}_t)_{t \in \{0,1,\dots,T\}}$ in the following way

$$\tilde{S}_t = S_0 A_t N_t$$

where A_t is a predictable process with respect to $\mathcal{G}_t$ and N_t is a $\mathcal{G}_t$ martingale under $\mathbb{P}$. From

$$\frac{\tilde{S}_{t+1}}{\tilde{S}_t} = \frac{A_{t+1} N_{t+1}}{A_t N_t}$$

we deduce that

$$E_{\mathbb{P}}\left[\frac{\tilde{S}_{t+1}}{\tilde{S}_t} \mid \mathcal{G}_t\right] = \frac{A_{t+1}}{A_t},$$

thus, A_t is uniquely defined by

$$A_t = \prod_{i=0}^{t-1} E_{\mathbb{P}}\left[\frac{\tilde{S}_{i+1}}{\tilde{S}_i} \mid \mathcal{G}_i\right].$$

Under the historical probability $\mathbb{P}$, the dynamics of the risky asset is given by

$$\tilde{S}_t = \tilde{S}_{t-1} e^{\nu_t} \frac{N_t}{N_{t-1}} \tag{3.40}$$

where ν_t is the one-period discounted excess returns between $t-1$ and t:

$$e^{\nu_t} = \frac{A_t}{A_{t-1}} = e^{-r} E_{\mathbb{P}}\left[e^{Y_t} \mid \mathcal{G}_{t-1}\right]. \tag{3.41}$$

Let us denote by $g_t^{\mathbb{P}}$ the conditional density function of $\frac{N_t}{N_{t-1}}$ given $\mathcal{G}_{t-1}$ under $\mathbb{P}$ and by $(\overline{L}_t)_{t\in\{1,\dots,T\}}$ the process defined by

$$\overline{L}_t = \prod_{i=1}^{t} \frac{g_i^{\mathbb{P}}\left(\frac{\tilde{S}_i}{\tilde{S}_{i-1}}\right) e^{\nu_i}}{g_i^{\mathbb{P}}\left(e^{-\nu_i}\frac{\tilde{S}_i}{\tilde{S}_{i-1}}\right)} \tag{3.42}$$

that is a $\mathcal{G}_t$ martingale under $\mathbb{P}$. We have the following proposition that corresponds to a conditional version of Lemma 3.3.1 taking $e^Z = \frac{\tilde{S}_t}{\tilde{S}_{t-1}}$ and $m = \nu_t$:

Proposition 3.3.1 (Elliott and Madan 1998) *Let $\mathbb{Q}^{em}$ be the probability defined by the density $\overline{L}_T$ with respect to $\mathbb{P}$, then, $\mathbb{Q}^{em}$ is the unique probability such that, $\forall t \in \{1,\dots,T\}$, the conditional distribution of $\frac{\tilde{S}_t}{\tilde{S}_{t-1}}$ given $\mathcal{G}_{t-1}$ under $\mathbb{Q}^{em}$ is equal to the conditional distribution of $\frac{N_t}{N_{t-1}}$ given $\mathcal{G}_{t-1}$ under $\mathbb{P}$. In particular, $\mathbb{Q}^{em}$ is an equivalent martingale measure.*

In the GARCH setting, we simply have $\frac{\tilde{S}_t}{\tilde{S}_{t-1}} = e^{-r+Y_t}$ and $\frac{N_t}{N_{t-1}} = e^{-r+Y_t-\nu_t}$. Using Lemma 3.3.1, is it possible to find explicitly a unique probability on $(\Omega, \mathcal{G}_t)$ under which the conditional distribution of e^{-r+Y_t} given $\mathcal{G}_{t-1}$ is equal to the conditional distribution of $e^{-r+Y_t-\nu_t}$ given $\mathcal{G}_{t-1}$. The definition of $\mathbb{Q}^{em}$ and the proof of the preceding proposition become a consequence of Eq. (3.39).

Example 3.3.1 When the dynamics of Y_t is given by Eq. (3.4) with Gaussian innovations, we obtain from (3.41) that $\nu_t = m_t + \frac{h_t}{2}$ and the conditional distribution of Y_t given $\mathcal{G}_{t-1}$ is, under $\mathbb{Q}^{em}$, a $\mathcal{N}(r - \frac{h_t}{2}, h_t)$ distribution. We deduce, in the Gaussian GARCH setting, that $\mathbb{Q}^{em}$ fulfills the LRNVR of Duan (1995) implying a risk-neutral dynamics given by Proposition 3.2.2.

Although the extended Girsanov principle, as the LRNVR, may be justified by equilibrium arguments in the Gaussian case, it is not possible for other distributions (as remarked in Schroder 2004 and Badescu et al. 2009) to obtain the analogous of Proposition 3.2.3. In fact, contrary to Eq. (3.29), it is difficult to extract from the equality

$$\frac{g_t^{\mathbb{P}}\left(\frac{\tilde{S}_t}{\tilde{S}_{t-1}}\right) e^{\nu_t}}{g_t^{\mathbb{P}}\left(e^{-\nu_t}\frac{\tilde{S}_t}{\tilde{S}_{t-1}}\right)} = e^{r-\rho}\frac{u'(c_t)}{u'(c_{t-1})}$$

a simple relation between the aggregate consumption c_t and the log-returns Y_t. Nevertheless, Elliott and Madan (1998) showed that the preceding specification of an EMM using the extended Girsanov principle may be economically justified considering hedging strategies that minimize the conditional variance of the discounted, risk-adjusted, hedging cost in the spirit of Föllmer and Schweiser (1991).

3.3.2 Risk-Neutral Dynamics for Classical Distributions

Due to the simplicity of the change in distribution implied by the EMM $\mathbb{Q}^{em}$, several authors have studied the associated risk-neutral dynamics for particular choices of distributions for the innovations (see for instance Menn and Rachev 2005 for the α-stable case, Badescu et al. 2008 for the MN distribution and Badescu et al. 2011 for the GH one). Here, we are able to obtain an easy and very general result that nests the preceding examples.

If $x = [x_i]_{1\leq i\leq n}$ is a vector of $\mathbb{R}^n$ with $n \in \mathbb{N}^*$ and if M is a real number we denote by $M + x$ the vector $[x_i + M]_{1\leq i\leq n}$. For $(p, q) \in (\mathbb{N}^*)^2$, we consider a family $(D(\theta, \overline{\theta}))_{(\theta,\overline{\theta})(\Theta,\overline{\Theta})}$ of distributions on $\mathbb{R}$ where $\Theta \subset \mathbb{R}^p$ and $\overline{\Theta} \subset \mathbb{R}^q$ are two open sets. We suppose that there exist mappings $(\Psi_\Sigma)_{\Sigma\in\mathbb{R}^*}$ from $\overline{\Theta}$ into $\overline{\Theta}$ such that $\forall(M, \Sigma) \in \mathbb{R} \times \mathbb{R}^*$,

$$M + \Sigma D(\theta, \overline{\theta}) = D(M + \Sigma\theta, \Psi_\Sigma(\overline{\theta})).$$

In particular, θ is the location parameter.

Remark 3.3.1 The preceding hypothesis is fulfilled for the distributions considered in Sect. 2.4. In the case of a $GH(\lambda, \alpha, \beta, \delta, \mu)$ we can take $\theta = \mu$, $\overline{\theta} = (\lambda, \alpha, \beta, \delta)$ and

$$\Psi_\Sigma(\lambda, \alpha, \beta, \delta) = (\lambda, \frac{\alpha}{\Sigma}, \frac{\beta}{\Sigma}, \Sigma\delta)$$

and in the case of a $MN(\phi, \mu_1, \mu_2, \sigma_1, \sigma_2)$, $\theta = (\mu_1, \mu_2)$, $\overline{\theta} = (\phi, \sigma_1, \sigma_2)$ and

$$\Psi_\Sigma(\phi, \sigma_1, \sigma_2) = (\phi, \Sigma\sigma_1, \Sigma\sigma_2).$$

We have the following result that is a consequence of Proposition 3.3.1:

Proposition 3.3.2 *When the dynamics of the risky asset under* $\mathbb{P}$ *is given by Eqs.* (3.4) *and* (3.5) *where the* z_t*'s are i.i.d* $D(\theta, \overline{\theta})$ *(with mean* 0 *and variance* 1*) then, under* $\mathbb{Q}^{em}$*, the distribution of* Y_t *given* $\mathcal{F}_{t-1}$ *is a*

$$D(m_t + r - \nu_t + \sqrt{h_t}\theta, \Psi_{\sqrt{h_t}}(\overline{\theta})) \tag{3.43}$$

where ν_t *is defined by Eq.* (3.41)*. Thus, under* $\mathbb{Q}^{em}$*,*

$$Y_t = r + m_t - \nu_t + \sqrt{h_t}\xi_t, \quad S_0 = s,$$

$$h_t = F(\xi_{t-1} - \frac{\nu_{t-1}}{\sqrt{h_{t-1}}}, h_{t-1})$$

where the ξ_t *are, under* $\mathbb{Q}^{em}$*, i.i.d random variables following a* $D(\theta, \overline{\theta})$*.*

Corollary 3.3.1 *For classical parametric cases, we are able to give explicitly the risk-neutral dynamics under* $\mathbb{Q}^{em}$*:*

1. *When* $D(\theta,\overline{\theta}) = GH(\lambda,\alpha,\beta,\delta,\mu)$*, the dynamics of the risky asset under* $\mathbb{Q}^{em}$ *is given by the preceding proposition with*

$$e^{\nu_t} = e^{\mu\sqrt{h_t}+m_t}\left(\frac{\alpha^2-\beta^2}{\alpha^2-(\beta+\sqrt{h_t})^2}\right)^{\frac{\lambda}{2}}\frac{K_\lambda(\delta\sqrt{\alpha^2-(\beta+\sqrt{h_t})^2})}{K_\lambda(\delta\sqrt{\alpha^2-\beta^2})}.$$

2. *When* $D(\theta,\overline{\theta}) = MN(\phi,\mu_1,\mu_2,\sigma_1,\sigma_2)$*, the dynamics of the risky asset under* $\mathbb{Q}^{em}$ *is given by the preceding proposition with*

$$e^{\nu_t} = e^{m_t}\left(\phi e^{\sqrt{h_t}\mu_1+\frac{\sigma_1^2 h_t}{2}} + (1-\phi)e^{\sqrt{h_t}\mu_2+\frac{\sigma_2^2 h_t}{2}}\right).$$

One of the main advantage of the extended Girsanov principle with respect to other SDF specifications comes from Eq. (3.41) which can be solved explicitly without approximation algorithms while in general we need to solve numerically the martingale equations. When we use Monte Carlo methods to approximate options prices, the running time is thereby reduced drastically. Nevertheless, it is important to remind that from $\mathbb{P}$ to $\mathbb{Q}^{em}$ the only change is a shift in the conditional distributions, in particular, the conditional variance, the conditional skewness and the conditional kurtosis of the innovations remain constant. It may explain partly the poor pricing performances of this method for long maturity options as remarked in Badescu and Kulperger (2008) and Badescu et al. (2011). The next method overcomes this empirical problem.

3.4 The Conditional Esscher Transform

3.4.1 Definition and Properties

In the seminal paper of Gerber and Shiu (1994a), an elegant way to choose an equivalent martingale measure in a continuous time and Markovian setting is provided using the so-called Esscher transform. This tool was first introduced in actuarial science by Esscher (1932). In contrast to Duan's (1995) approach, this latter framework permits a wide variety of return innovations to be chosen within the class of infinite divisible distributions. This method has been adapted in Bühlmann et al. (1996) to price options in discrete time financial models and first applied by Siu et al. (2004) (see also Christoffersen et al. 2010) in the GARCH setting with possible extensions to Markow switching (Elliott et al. 2006), multi-component (Christoffersen et al. 2008) and multiple-shock GARCH models (Christoffersen et al. 2012). An equivalent formulation of the work of Gerber and Shiu (1994a) consists in using an exponential affine parameterization for the SDF (see Gouriéroux and Monfort 2007) that will be adopted in this section.[9]

The methodology unfolds as follows. We assume for the SDF a particular parametric form (exponential affine of the log-returns): $\forall t \in \{1, \ldots, T\}$,

$$M_t = e^{\theta_t Y_t + \xi_t} \tag{3.44}$$

where $Y_t = log\left(\frac{S_t}{S_{t-1}}\right)$ and where θ_t and ξ_t are $\mathcal{F}_{t-1}$ measurable random variables.

Remark 3.4.1 From Remark 3.2.4, the parameter $\theta_t = \frac{\partial log(M_t)}{\partial Y_t}$ corresponds to the opposite of the absolute risk aversion coefficient and will be in general strictly negative. We have already seen (Eq. (3.20)) that, in a discrete time version of the Black and Scholes economy, the corresponding SDF is of the form (3.44) with $\theta_t = \frac{r-\mu}{\sigma^2}$ and $\xi_t = \frac{(r+\mu-\sigma^2)(\mu-r)}{2\sigma^2} - r$ that are independent of t. Here, the general expression (3.44) allows for predictable time variations in risk aversion.

We need to compute explicitly $(\theta_{t+1}, \xi_{t+1})$. Considering the bond and the risky asset, the pricing relations (3.14) and (3.15) give the following restrictions for the parameters

$$\begin{cases} E_{\mathbb{P}}[e^r M_{t+1} \mid \mathcal{F}_t] = 1 \\ E_{\mathbb{P}}[e^{Y_{t+1}} M_{t+1} \mid \mathcal{F}_t] = 1. \end{cases} \tag{3.45}$$

For all $t \in \{0, \ldots, T-1\}$, we denote by $\mathbb{G}^{\mathbb{P}}_{Y_{t+1}|\mathcal{F}_t}$ the conditional moment generating function, under $\mathbb{P}$, of Y_{t+1} given $\mathcal{F}_t$:

$$\mathbb{G}^{\mathbb{P}}_{Y_{t+1}|\mathcal{F}_t}(u) = E_{\mathbb{P}}\left[e^{uY_{t+1}} \mid \mathcal{F}_t\right],$$

defined on a convex set $\mathcal{D}_{\mathbb{G}^{\mathbb{P}}_{Y_{t+1}|\mathcal{F}_t}}$ which is not reduced to $\{0\}$, and by Θ_t the set of parameters

$$\left\{\theta \in \mathbb{R}; \theta \text{ and } 1+\theta \in \mathcal{D}_{\mathbb{G}^{\mathbb{P}}_{Y_{t+1}|\mathcal{F}_t}}\right\}.$$

We now introduce the mapping $\Phi_t : \Theta_t \to \mathbb{R}$ defined by

$$\Phi_t(\theta) = log\left(\frac{\mathbb{G}^{\mathbb{P}}_{Y_{t+1}|\mathcal{F}_t}(1+\theta)}{\mathbb{G}^{\mathbb{P}}_{Y_{t+1}|\mathcal{F}_t}(\theta)}\right).$$

[9] As remarked in Heston and Nandi (2000) for Gaussian innovations, this exponential affine parameterization is equivalent to obtaining of Black and Scholes prices for call options with 1 day to expiration.

Thus, the system (3.45) is equivalent to

$$\begin{cases} \mathbb{G}^{\mathbb{P}}_{Y_{t+1}|\mathcal{F}_t}(\theta_{t+1}) = e^{-(r+\xi_{t+1})} \\ \mathbb{G}^{\mathbb{P}}_{Y_{t+1}|\mathcal{F}_t}(\theta_{t+1}+1) = e^{-\xi_{t+1}} \end{cases} \tag{3.46}$$

and, with our notations, we have to solve

$$\begin{cases} \Phi_t(\theta_{t+1}) = r \\ \mathbb{G}^{\mathbb{P}}_{Y_{t+1}|\mathcal{F}_t}(\theta_{t+1}+1) = e^{-\xi_{t+1}}. \end{cases} \tag{3.47}$$

The next proposition shows that, under the pricing constraints (3.45), there is no ambiguity in the choice of the SDF (3.44).

Proposition 3.4.1 (Gerber and Shiu 1994b) *Suppose that* $\mathbb{G}^{\mathbb{P}}_{Y_{t+1}|\mathcal{F}_t}$ *is twice differentiable on* $\mathcal{D}_{\mathbb{G}^{\mathbb{P}}_{Y_{t+1}|\mathcal{F}_t}}$. *If there exists a solution to the equation* $\Phi_t(\theta) = r$, *it is unique.*

Subsequently, we suppose for each $t \in \{0, \ldots, T-1\}$ that (3.47) leads to a unique solution denoted by $(\theta^q_{t+1}, \xi^q_{t+1})$ [general sufficient conditions may be found in Christoffersen et al. (2010) for the existence].[10] From (3.19), we deduce the form of the associated EMM denoted $\mathbb{Q}^{ess}$:

$$\frac{d\mathbb{Q}^{ess}}{d\mathbb{P}} = e^{rT}\prod_{k=1}^{T} M_k = \prod_{k=1}^{T} \frac{e^{\theta^q_k Y_k}}{\mathbb{G}^{\mathbb{P}}_{Y_k|\mathcal{F}_{k-1}}(\theta^q_k)} \tag{3.48}$$

that is none other than the conditional Esscher transform of parameters $(\theta^q_k)_{k\in\{1,\ldots,T\}}$ of the historical probability $\mathbb{P}$ (see Bühlmann et al. 1996). The next proposition, that is a direct consequence of the Proposition A.4 of the Appendix, describes the dynamics of the risky asset under $\mathbb{Q}^{ess}$:

Proposition 3.4.2 *Under* $\mathbb{Q}^{ess}$, *the moment generating function of* Y_t *given* $\mathcal{F}_{t-1}$ *is given by*

$$\mathbb{G}^{\mathbb{Q}^{ess}}_{Y_t|\mathcal{F}_{t-1}}(u) = E_{\mathbb{P}}\left[e^{uY_t}\frac{e^{\theta^q_t Y_t}}{\mathbb{G}^{\mathbb{P}}_{Y_t|\mathcal{F}_{t-1}}(\theta^q_t)} \mid \mathcal{F}_{t-1}\right] = \frac{\mathbb{G}^{\mathbb{P}}_{Y_t|\mathcal{F}_{t-1}}(\theta^q_t+u)}{\mathbb{G}^{\mathbb{P}}_{Y_t|\mathcal{F}_{t-1}}(\theta^q_t)}. \tag{3.49}$$

[10] We will elaborate on this point later on for particular distributions using the intermediate value theorem (see Propositions 3.4.5 and 3.4.8).

Following Christoffersen et al. (2010), we are able to characterize the conditional risk-neutral variance and skewness from the preceding proposition.[11] We denote by $\Psi^{\mathbb{P}}_{\varepsilon_t}$ (resp. $\Psi^{\mathbb{Q}^{ess}}_{\varepsilon_t}$) the logarithm of the moment generating function (supposed to be regular) of the innovation ε_t given $\mathcal{F}_{t-1}$ under $\mathbb{P}$ (resp. $\mathbb{Q}^{ess}$). From Eq. (3.49)

$$\Psi^{\mathbb{Q}^{ess}}_{\varepsilon_t}(u) = \Psi^{\mathbb{P}}_{\varepsilon_t}(u+\theta^q_t) - \Psi^{\mathbb{P}}_{\varepsilon_t}(\theta^q_t),$$

thus,

$$E_{\mathbb{Q}^{ess}}[\varepsilon_t \mid \mathcal{F}_{t-1}] = \left(\Psi^{\mathbb{Q}^{ess}}_{\varepsilon_t}\right)'(0) = \left(\Psi^{\mathbb{P}}_{\varepsilon_t}\right)'(\theta^q_t)$$

and we define the risk-neutral innovation $\varepsilon^*_t = \varepsilon_t - \left(\Psi^{\mathbb{P}}_{\varepsilon_t}\right)'(\theta^q_t)$. We have in particular

$$\Psi^{\mathbb{Q}^{ess}}_{\varepsilon^*_t}(u) = \Psi^{\mathbb{Q}^{ess}}_{\varepsilon_t}(u) - u\left(\Psi^{\mathbb{P}}_{\varepsilon_t}\right)'(\theta^q_t)$$

and the conditional variance (h^*_t) and skewness (s^*_t) of ε^*_t under $\mathbb{Q}^{ess}$ are equal to

$$h^*_t = \left(\Psi^{\mathbb{P}}_{\varepsilon_t}\right)''(\theta^q_t) = h_t \frac{\left(\Psi^{\mathbb{P}}_{\varepsilon_t}\right)''(\theta^q_t)}{\left(\Psi^{\mathbb{P}}_{\varepsilon_t}\right)''(0)}$$

$$s^*_t = \frac{\left(\Psi^{\mathbb{P}}_{\varepsilon_t}\right)'''(\theta^q_t)}{(h^*_t)^{\frac{3}{2}}} = s_t \frac{\left(\Psi^{\mathbb{P}}_{\varepsilon_t}\right)'''(\theta^q_t)}{\left(\Psi^{\mathbb{P}}_{\varepsilon_t}\right)'''(0)} \left(\frac{h_t}{h^*_t}\right)^{\frac{3}{2}}.$$

Supposing that the following Taylor approximations hold

$$\begin{cases} \left(\Psi^{\mathbb{P}}_{\varepsilon_t}\right)''(\theta^q_t) \approx \left(\Psi^{\mathbb{P}}_{\varepsilon_t}\right)''(0) + \left(\Psi^{\mathbb{P}}_{\varepsilon_t}\right)'''(0)\theta^q_t + \left(\Psi^{\mathbb{P}}_{\varepsilon_t}\right)''''(0)\frac{(\theta^q_t)^2}{2} \\ \left(\Psi^{\mathbb{P}}_{\varepsilon_t}\right)'''(\theta^q_t) \approx \left(\Psi^{\mathbb{P}}_{\varepsilon_t}\right)'''(0) + \left(\Psi^{\mathbb{P}}_{\varepsilon_t}\right)''''(0)\theta^q_t \end{cases}$$

we finally obtain

$$\begin{cases} h^*_t \approx h_t + s_t(h_t)^{\frac{3}{2}}\theta^q_t + \frac{(\theta^q_t)^2}{2}h^2_t k_t \\ s^*_t \approx s_t\left(\frac{h_t}{h^*_t}\right)^{\frac{3}{2}} + k_t \frac{h^2_t}{(h^*_t)^{\frac{3}{2}}}\theta^q_t \end{cases}$$

[11] When X is a centered random variable such that the mapping $\Psi(u) = log(E[e^{uX}])$ is regular we have $\Psi'(0) = E[X] = 0$, $\Psi''(0) = Var[X]$, $\frac{\Psi'''(0)}{(\Psi''(0))^{\frac{3}{2}}} = sk[X]$ and $\frac{\Psi''''(0)}{(\Psi''(0))^2} = k[X] - 3$.

where k_t is the conditional excess kurtosis of ε_t under $\mathbb{P}$. From the preceding approximations, it is interesting to remark that in the realistic case where $s_t < 0$ and $k_t > 0$ (see Table 2.5) we have $h_t^* > h_t$. In fact, we will find empirically that the parameter θ_t^q is in general strictly negative.[12] As a consequence, when $s_t < 0$ and $k_t > 0$ then $s_t^* < s_t$. Thus, starting from realistic distributions for the innovations, the Esscher option pricing framework is able to cope with the well-known stylized facts that the risk-neutral variance is greater than the historical one and that the risk-neutral distribution of returns is more negatively skewed than the physical distribution. In the opposite case, when $s_t = k_t = 0$ we have $h_t = h_t^*$ and $s_t^* = 0$. In particular, for Gaussian innovations the following risk-neutral dynamics is deduced:

Example 3.4.1 In the Gaussian case, the historical conditional distribution of Y_t given $\mathcal{F}_{t-1}$ is a $\mathcal{N}(r + m_t, h_t)$ and we deduce from (3.47) that $\theta_t^q = -(\frac{1}{2} + \frac{m_t}{h_t})$. Thus, from Proposition 3.4.2,

$$\mathbb{G}^{\mathbb{Q}^{ess}}_{Y_t|\mathcal{F}_{t-1}}(u) = e^{u(r-\frac{h_t}{2})+\frac{u^2 h_t}{2}}$$

and we recover once again the risk-neutral dynamics coming from the LRNVR in Proposition 3.2.2. In particular, having Proposition 3.2.3 in mind, the conditional Esscher transform is consistent with equilibrium pricing models for Gaussian residuals. In the next proposition we see that this consistency may be extended from Gaussian to general distributions.

Proposition 3.4.3 (Badescu et al. 2009) *In the framework of Sect. 3.2.3, when the dynamics of the log-returns is given by Eq.* (3.4) *where z_t follows a centered distribution $\mathcal{D}(0, 1)$ with variance one, the equilibrium EMM coincides with $\mathbb{Q}^{ess}$ under any of the three following conditions:*

(1) u is linear.
(2) The utility function has the form (3.28) *and*

$$\Delta log(c_t) = log(c_t) - log(c_{t-1}) = A_t^0 + A_t^1 Y_t + A_t^2 \eta_t$$

where the processes $(A_t^0)_{t\in\{1,\dots,T\}}$, $(A_t^1)_{t\in\{1,\dots,T\}}$ and $(A_t^2)_{t\in\{1,\dots,T\}}$ are predictable and where $(\eta_t)_{t\in\{1,\dots,T\}}$ is a sequence of i.i.d $\mathcal{D}(0, 1)$ such that η_t is independent of z_t given $\mathcal{F}_{t-1}$.

[12] See Example 3.4.1 or Remark 3.4.1 for an intuitive interpretation of this modified Sharpe ratio in terms of risk aversion.

(3) *The utility function has an exponential form* $u(x) = \frac{1-e^{-Rx}}{R}$ *where* $R > 0$ *is the coefficient of absolute risk aversion and*

$$\Delta(c_t) = c_t - c_{t-1} = A_t^0 + A_t^1 Y_t + A_t^2 \eta_t$$

where the processes $(A_t^0)_{t\in\{1,\dots,T\}}$, $(A_t^1)_{t\in\{1,\dots,T\}}$ *and* $(A_t^2)_{t\in\{1,\dots,T\}}$ *are predictable and where* $(\eta_t)_{t\in\{1,\dots,T\}}$ *is a sequence of i.i.d* $\mathcal{D}(0,1)$ *such that* η_t *is independent of* z_t *given* $\mathcal{F}_{t-1}$.

In the two last points, we have $\theta_t^q = -RA_t^1$ *proving that in general* $\theta_t^q < 0$ *because the correlation coefficient* A_t^1 *is in general strictly positive as remarked in Breeden et al. (1989).*

Remark 3.4.2 The proof of this proposition follows the same steps as in Sect. 3.2.3 and is omitted. We refer the reader to Badescu et al. (2009) for details. As in the Gaussian case (Proposition 3.2.3), it is sometimes possible (see Badescu et al. 2008, 2011) to relax the regression conditions for the aggregate consumption imposing a particular bivariate distribution for the pairs $(Y_t, \Delta log(c_t))$ or $(Y_t, \Delta(c_t))$.

To deal with practical issues, it is important to fully characterized the risk-neutral dynamics of the risky asset. As underlined in the next proposition, under weak assumptions, the conditional distribution of the risky asset under $\mathbb{Q}^{ess}$ belongs to the same family as under the historical probability.

Proposition 3.4.4 (Chorro et al. 2012)

(1) *For all* $t \in \{1,\dots,T\}$, *if the conditional distribution of* Y_t *given* $\mathcal{F}_{t-1}$ *is infinitely divisible under* $\mathbb{P}$ *and if* $\mathbb{G}^{\mathbb{P}}_{Y_t|\mathcal{F}_{t-1}}$ *is twice differentiable, then, the conditional distribution of* Y_t *given* $\mathcal{F}_{t-1}$ *is also infinitely divisible under* $\mathbb{Q}^{ess}$ *with a finite moment of order 2.*
(2) *For all* $t \in \{1,\dots,T\}$, *if the conditional distribution of* Y_t *given* $\mathcal{F}_{t-1}$ *is a finite mixture under* $\mathbb{P}$ *the same property holds under* $\mathbb{Q}^{ess}$.

In the next section we improve the preceding result showing that for particular sub-families of distributions the change of parameters from $\mathbb{P}$ to $\mathbb{Q}^{ess}$ may be perfectly described.

3.4.2 Risk-Neutral Dynamics for Classical Distributions

In this section, we consider, among other things, for the innovations of (3.4) the probability distributions introduced in Sect. 2.4.

The Generalized Hyperbolic Distribution

For $(\lambda, \alpha, \beta, \delta, \mu) \in \mathbb{R}^5$ with $\delta > 0$ and $\alpha >| \beta |> 0$, we suppose that the z_t are i.i.d random variables following a one dimensional $GH(\lambda, \alpha, \beta, \delta, \mu)$ distribution that is defined in Sect. 2.4.1 by the following density function:

$$d_{GH}(x, \lambda, \alpha, \beta, \delta, \mu) = \frac{(\sqrt{\alpha^2 - \beta^2}/\delta)^\lambda}{\sqrt{2\pi} K_\lambda(\delta \sqrt{\alpha^2 - \beta^2})} e^{\beta(x-\mu)} \frac{K_{\lambda - 1/2}\left(\alpha \sqrt{\delta^2 + (x-\mu)^2}\right)}{\left(\sqrt{\delta^2 + (x-\mu)^2}/\alpha\right)^{1/2-\lambda}} \tag{3.50}$$

with an associated moment generating function given by:

$$\mathbb{G}_{GH}(u) = e^{\mu u} \left(\frac{\alpha^2 - \beta^2}{\alpha^2 - (\beta + u)^2} \right)^{\frac{\lambda}{2}} \frac{K_\lambda(\delta \sqrt{\alpha^2 - (\beta + u)^2})}{K_\lambda(\delta \sqrt{\alpha^2 - \beta^2})}, \quad | \beta + u |< \alpha. \tag{3.51}$$

Remark 3.4.3 To standardize the distribution we implicitly suppose that the parameters $(\lambda, \alpha, \beta, \delta, \mu)$ satisfy the following conditions

$$\begin{cases} E_{\mathbb{P}}[z_t] = \mu + \frac{\delta\beta}{\gamma} \frac{K_{\lambda+1}(\delta\gamma)}{K_\lambda(\delta\gamma)} = 0 \\ Var_{\mathbb{P}}[z_t] = \frac{\delta}{\gamma} \frac{K_{\lambda+1}(\delta\gamma)}{K_\lambda(\delta\gamma)} + \frac{\beta^2\delta^2}{\gamma^2} \left(\frac{K_{\lambda+2}(\delta\gamma)K_\lambda(\delta\gamma) - K^2_{\lambda+1}(\delta\gamma)}{K^2_\lambda(\delta\gamma)} \right) = 1. \end{cases} \tag{3.52}$$

where $\gamma = \sqrt{\alpha^2 - \beta^2}$.[13]

From the stability of the GH distribution under affine transformations (Barndorff-Nielsen and Blaesild 1981) we deduce that, under $\mathbb{P}$, the conditional distribution of Y_t given $\mathcal{F}_{t-1}$ is a

$$GH\left(\lambda, \frac{\alpha}{\sqrt{h_t}}, \frac{\beta}{\sqrt{h_t}}, \delta\sqrt{h_t}, r + m_t + \mu\sqrt{h_t}\right)$$

and so for $| \beta + u\sqrt{h_t} |< \alpha$,

$$\mathbb{G}^{\mathbb{P}}_{Y_t|\mathcal{F}_{t-1}}(u) = e^{u(\mu\sqrt{h_t}+r+m_t)} \left(\frac{\alpha^2 - \beta^2}{\alpha^2 - (\beta + u\sqrt{h_t})^2} \right)^{\frac{\lambda}{2}} \frac{K_\lambda\left(\delta \sqrt{\alpha^2 - (\beta + u\sqrt{h_t})^2}\right)}{K_\lambda(\delta \sqrt{\alpha^2 - \beta^2})}. \tag{3.53}$$

[13] Sometimes it is interesting to work with an unconstrained GH distribution and it is sufficient to consider $\tilde{z}_t = \frac{z_t - E_{\mathbb{P}}[z_t]}{\sqrt{Var_{\mathbb{P}}[z_t]}}$ instead of z_t in (3.4). This is the point of view of Badescu et al. (2011). The only difference between the two approaches is a modification in the parameterization of the model.

Now, following Chorro et al. (2012) we apply the methodology of the preceding section in this setting. We first obtain a result that ensures, under weak conditions, the existence of a solution to Eq. (3.47).[14]

Proposition 3.4.5 *For a $GH(\lambda, \alpha, \beta, \delta, \mu)$ distribution with $\alpha > \frac{1}{2}$, then,*

(1) If $\lambda \geq 0$, the equation $log\left(\frac{\mathbb{G}_{GH}(1+\theta)}{\mathbb{G}_{GH}(\theta)}\right) = r$ has a unique solution.
(2) If $\lambda < 0$, the equation $log\left(\frac{\mathbb{G}_{GH}(1+\theta)}{\mathbb{G}_{GH}(\theta)}\right) = r$ has a unique solution if and only if $\mu - C < r < \mu + C$ where

$$C = log\left(\frac{\Gamma[-\lambda]}{2^{\lambda+1}}\right) - log\left(\frac{K_\lambda(\delta\sqrt{\alpha^2 - (\alpha-1)^2})}{[\delta\sqrt{\alpha^2 - (\alpha-1)^2}]^\lambda}\right).$$

The constant C is strictly positive because $\frac{d}{dx}\frac{K_\lambda(x)}{x^\lambda} = -\frac{K_{\lambda+1}(x)}{x^\lambda} < 0$.

The next proposition describes the dynamics of the risky asset under $\mathbb{Q}^{ess}$. It derives directly from the Proposition 3.4.2.

Proposition 3.4.6 *Under $\mathbb{Q}^{ess}$, the distribution of Y_t given $\mathcal{F}_{t-1}$ is a*

$$GH\left(\lambda, \frac{\alpha}{\sqrt{h_t}}, \frac{\beta}{\sqrt{h_t}} + \theta_t^q, \delta\sqrt{h_t}, r + m_t + \mu\sqrt{h_t}\right). \quad (3.54)$$

We deduce from the preceding result that, under $\mathbb{Q}^{ess}$,

$$Y_t = r + m_t + \underbrace{\sqrt{h_t}\xi_t}_{\varepsilon_t^*}, \quad S_0 = s, \quad (3.55)$$

where the ξ_t are $\mathcal{F}_t$ measurable random variables such that, conditionally to $\mathcal{F}_{t-1}$,[15]

$$\xi_t \hookrightarrow GH(\lambda, \alpha, \beta + \sqrt{h_t}\theta_t^q, \delta, \mu). \quad (3.56)$$

[14] In general, we are able to obtain an explicit solution only in particular cases (see Proposition 3.4.7 below). Nevertheless, we can compute it efficiently using refined bracketing methods.

[15] For Gaussian innovations (see Example 3.4.1) only the first order conditional moment is impacted by the measure change. Here, θ_t^q induces not only a shift in the conditional skewness of the GH distribution but also an excess kurtosis (exact values for the associated skewness and excess kurtosis are provided in Barndorff-Nielsen and Blaesild 1981).

In particular, the GH distribution is preserved after this measure change and we can still simulate the sample paths of the risky asset. Under $\mathbb{Q}^{ess}$, conditionally to $\mathcal{F}_{t-1}$, ε_t^* is no longer centered and its variance is not h_t but

$$Var[\varepsilon_t^*] = h_t \left(\frac{\delta K_{\lambda+1}(\delta\gamma_t)}{\gamma_t K_\lambda(\delta\gamma_t)} + \frac{(\beta + \sqrt{h_t}\theta_t^q)^2\delta^2}{\gamma_t^2} \left(\frac{K_{\lambda+2}(\delta\gamma_t)}{K_\lambda(\delta\gamma_t)} - \frac{K_{\lambda+1}^2(\delta\gamma_t)}{K_\lambda^2(\delta\gamma_t)} \right) \right),$$

where $\gamma_t = \sqrt{\alpha^2 - (\beta + \sqrt{h_t}\theta_t^q)^2}$. Thus, the GARCH structure of the volatility is modified in a non-linear way from $\mathbb{P}$ to $\mathbb{Q}^{ess}$.

Remark 3.4.4 It is possible to recover the result of Proposition 3.4.6 using the proof of Proposition 3.4.4 in Sect. 3.8 and keeping in mind that the Kolmogorov representation of the GH distribution is known explicitly (see for example Guégan et al. 2013 that use this approach in the case $\lambda = -\frac{1}{2}$). Proposition 3.4.6 may also be seen as the discrete time counterpart of the unconditional Esscher pricing formula for exponential Lévy motions in the continuous time modelling approach (see Eberlein and Prause 2002).

In the current literature, there are various interesting subclasses and limits of the GH distribution (see Table 2.6) that are frequently used for modelling financial returns (see for example Barndorff-Nielsen 1995 and Eberlein and Prause 2002). Using Proposition 3.4.6 we are able to recover risk-neutral dynamics of three particular models, namely, the Gaussian GARCH option pricing models of Duan (1995) and Heston and Nandi (2000), the Gamma GARCH option pricing model of Siu et al. (2004)[16] and the Normal Inverse Gaussian GARCH option pricing model (see e.g. Guégan et al. 2013).

First, we state a lemma that explicitly reminds how to obtain the two first distributions from the GH family.

Lemma 3.4.1 (Eberlein and Hammerstein 2003)

(1) When $\mu = \beta = 0$ and $\alpha, \delta \to +\infty$ with $\frac{\alpha}{\delta} \to 1$, the GH distribution converges in distribution toward a $\mathcal{N}(0, 1)$.

(2) When $\mu = 0$, $\lambda = a > 0$, $\alpha, \beta \to +\infty$, $\alpha\delta^2 \to 0$ and $\alpha - \beta = b > 0$, the GH distribution converges in distribution toward a Gamma $G_a(a, b)$ distribution.

[16] In this paper, the authors assume that in Eq. (3.4), $z_t = \frac{x_t - \frac{a}{b}}{\sqrt{\frac{a}{b^2}}}$ where x_t follows a $G_a(a, b)$ distribution having the density $\frac{b^a x^{a-1} e^{-bx}}{\Gamma(a)} 1_{\mathbb{R}_+}(x)$.

Then, when $\lambda = -\frac{1}{2}$ (NIG case) or when the parameters converge as in the preceding lemma, we obtain the following proposition[17]:

Proposition 3.4.7 *(1) When $\mu = \beta = 0$ and $\alpha, \delta \to +\infty$ with $\frac{\alpha}{\delta} \to 1$, passing to the limit in (3.54), we obtain the Duan (1995) risk-neutral dynamics.*

(2) When $\mu = 0$, $\lambda = a > 0$, $\alpha, \beta \to +\infty$, $\alpha\delta^2 \to 0$ and $\alpha - \beta = b > 0$, the GH GARCH option pricing model converges under $\mathbb{Q}^{ess}$ toward the Gamma risk-neutral dynamics introduced in Siu et al. (2004).

(3) When $\lambda = -\frac{1}{2}$, the conditional distribution of Y_t given $\mathcal{F}_{t-1}$ under $\mathbb{Q}^{ess}$ is a

$$NIG\left(\frac{\alpha}{\sqrt{h_t}}, \frac{\beta}{\sqrt{h_t}} + \theta_t^q, \delta\sqrt{h_t}, r + m_t + \mu\sqrt{h_t}\right) \tag{3.57}$$

where

$$\theta_t^q = \frac{-1}{2} - \frac{\alpha\beta\sqrt{\delta}}{\sqrt{h_t}\gamma^{\frac{3}{2}}} - \frac{1}{2}\sqrt{\frac{\left(\alpha m_t + \sqrt{\delta h_t}\beta\gamma\right)^2}{h_t\delta\gamma^3}\left(\frac{4\alpha^4\delta^2}{h_t\delta\gamma^3 + \left(\alpha m_t + \sqrt{\delta h_t}\beta\gamma\right)^2} - 1\right)}. \tag{3.58}$$

Mixture of Gaussian Distributions

To lighten the notations, we will restrict ourselves to the case of the mixture of two Gaussian distributions that depends on the same number of parameters as the GH family.[18] In Eq. (3.4), we suppose that z_t follows a $MN(\phi, \mu_1, \mu_2, \sigma_1, \sigma_2)$ distribution introduced in Sect. 2.4 with a moment generating function given by

$$\mathbb{G}_{MN}(u) = \phi e^{u\mu_1 + \frac{\sigma_1^2 u^2}{2}} + (1-\phi)e^{u\mu_2 + \frac{\sigma_2^2 u^2}{2}}. \tag{3.59}$$

This family being stable under affine transformations we deduce that under $\mathbb{P}$, the conditional distribution of Y_t given $\mathcal{F}_{t-1}$ is a

$$MN(\phi, r + m_t + \sqrt{h_t}\mu_1, r + m_t + \sqrt{h_t}\mu_2, \sqrt{h_t}\sigma_1, \sqrt{h_t}\sigma_2),$$

thus,

$$\mathbb{G}^{\mathbb{P}}_{Y_t|\mathcal{F}_{t-1}}(u) = \phi e^{u(r+m_t+\sqrt{h_t}\mu_1) + \frac{h_t\sigma_1^2 u^2}{2}} + (1-\phi)e^{u(r+m_t+\sqrt{h_t}\mu_2) + \frac{h_t\sigma_2^2 u^2}{2}}. \tag{3.60}$$

[17] In these three cases, θ_t^q has an analytic form: the pricing equations (3.47) have an explicit solution.

[18] Alexander and Lazar (2006) prove that there is no real forecasting improvement introducing more than two components in the mixture.

Remark 3.4.5 The GARCH dynamics implied by Eq. (3.60) is different from the multi-component normal mixture GARCH models studied in Alexander and Lazar (2006) and Badescu et al. (2008). In fact in our model the total variance of the residuals follows a GARCH dynamics while in these papers the conditional distribution of the residuals is a mixture of Gaussian distributions where each variance component follows its own GARCH updating process. Nevertheless, in spite of the differences in the specifications of the models, the pricing framework and the associated proofs are deeply similar (Badescu et al. 2008).

Proposition 3.4.8 *(1) For a* $MN(\phi, \mu_1, \mu_2, \sigma_1, \sigma_2)$*, the equation* $\log\left(\frac{\mathbb{G}_{MN}(1+\theta)}{\mathbb{G}_{MN}(\theta)}\right) = r$ *has a unique solution.*

(2) Under $\mathbb{Q}^{ess}$*, the distribution of* Y_t *given* $\mathcal{F}_{t-1}$ *is a*

$$MN(\overline{\phi_t}, \overline{\mu_{1,t}}, \overline{\mu_{2,t}}, \overline{\sigma_{1,t}}, \overline{\sigma_{2,t}}) \tag{3.61}$$

where $\overline{\sigma_{1,t}} = \sqrt{h_t}\sigma_1$, $\overline{\sigma_{2,t}} = \sqrt{h_t}\sigma_2$, $\overline{\mu_{1,t}} = r + m_t + \sqrt{h_t}\mu_1 + \overline{\sigma_{1,t}}^2\theta_t^q$, $\overline{\mu_{2,t}} = r + m_t + \sqrt{h_t}\mu_2 + \overline{\sigma_{2,t}}^2\theta_t^q$*, and*

$$\overline{\phi_t} = \frac{\phi e^{\theta_t^q(r+m_t+\sqrt{h_t}\mu_1)+\overline{\sigma_{1,t}}^2\frac{(\theta_t^q)^2}{2}}}{\phi e^{\theta_t^q(r+m_t+\sqrt{h_t}\mu_1)+\overline{\sigma_{1,t}}^2\frac{(\theta_t^q)^2}{2}} + (1-\phi)e^{\theta_t^q(r+m_t+\sqrt{h_t}\mu_2)+\overline{\sigma_{2,t}}^2\frac{(\theta_t^q)^2}{2}}}.$$

Gaussian Jumps

In Duan et al. (2005, 2006) the authors introduce non-normality in classical GARCH type models with Gaussian innovations adding Gaussian jumps with a number of jumps between two time units characterized by a Poisson distribution $\mathcal{P}(\lambda)$ (see the Mathematical Appendix). More specifically, they consider the following modelling for the log-returns:

$$Y_t = \log\left(\frac{S_t}{S_{t-1}}\right) = r + m_t + \sqrt{h_t}\underbrace{\left(z_t^0 + \sum_{i=1}^{N_t} z_t^i\right)}_{z_t}, \quad S_0 = s \in \mathbb{R}_+, \tag{3.62}$$

$$h_t = F(z_{t-1}, h_{t-1}) \tag{3.63}$$

where $\left(z_t^0, N_t, z_t^i\right)_{i\in\mathbb{N}^*, t\in\{1,\dots,T\}}$ are independent random variables such that z_t^0 follows a $\mathcal{N}(0,1)$, N_t follows a $\mathcal{P}(\lambda)$ and z_t^i follows a $\mathcal{N}(\mu, \sigma^2)$.[19] In this case,

[19]First, in this model, the Gaussian and the jump components are intrinsically linked: they have up to a constant the same conditional variance because

$$Var_{\mathbb{P}}\left[\sqrt{h_t}z_t^0 \mid \mathcal{F}_{t-1}\right] = h_t \text{ and } Var_{\mathbb{P}}\left[\sqrt{h_t}\sum_{i=1}^{N_t} z_t^i \mid \mathcal{F}_{t-1}\right] = h_t\lambda(\mu^2 + \sigma^2).$$

the moment generating function of the log-returns is

$$\mathbb{G}^{\mathbb{P}}_{Y_t|\mathcal{F}_{t-1}}(u) = e^{\left[(r+m_t)u+\frac{u^2h_t}{2}+\lambda\left(e^{\mu\sqrt{h_t}u+\frac{u^2h_t\sigma^2}{2}}-1\right)\right]}$$

and the dynamics under the Esscher EMM is given by the following proposition:

Proposition 3.4.9 *Under $\mathbb{Q}^{ess}$, we have*

$$Y_t = log\left(\frac{S_t}{S_{t-1}}\right) = r + m_t + \theta_t^q h_t + \sqrt{h_t}\underbrace{\left(\tilde{z}_t^0 + \sum_{i=1}^{\tilde{N}_t} \tilde{z}_t^i\right)}_{\tilde{z}_t}, \quad S_0 = s \in \mathbb{R}_+, \tag{3.64}$$

$$h_t = F(\tilde{z}_{t-1} + \theta_{t-1}^q \sqrt{h_{t-1}}, h_{t-1}) \tag{3.65}$$

where θ_t^q is the unique solution of Eq. (3.47) *and where $\forall t \in \{1,\ldots,T\}$ the random variables $(\tilde{z}_t^0, \tilde{N}_t, \tilde{z}_t^i)_{i\in\mathbb{N}^*}$ are independent and such that $\tilde{z}_t^0$ follows a $\mathcal{N}(0,1)$, $\tilde{z}_t^i$ follows a $\mathcal{N}(\mu + \theta_t^q\sqrt{h_t}\sigma^2, \sigma^2)$ and $\tilde{N}_t$ follows a $\mathcal{P}(\lambda e^{\mu\theta_t^q\sqrt{h_t}+\frac{(\theta_t^q)^2h_t\sigma^2}{2}})$.*

As seen in the three preceding examples, the exponential affine parameterization of the stochastic discount factor appears as a flexible way to obtain explicit risk-neutral dynamics under very general assumptions for the distribution of the residuals. Moreover, the corresponding conditional variances and skewness under $\mathbb{Q}^{ess}$ are compatible with empirical studies but only in the case of heavy tailed and skewed innovations.

In the next subsection, we are going to see how to modify the exponential affine Esscher SDF in order to cope with these two empirical features only working with the simple Gaussian distribution.

3.5 Second Order Esscher Transform

When we work with Gaussian innovations, Sects. 3.3 and 3.4 showed that the Elliot and Madan (Example 3.3.1) and the Esscher (Example 3.4.1) EMM give rise to risk-neutral dynamics that are consistent with the Duan (1995) LRNVR. In particular,

---

We refer the reader to Christoffersen et al. (2012) for a multiple-shock version of the model where the Gaussian and the jump parts have distinct volatility dynamics. Finally, in Duan et al. (2005), the excess return m_t has a very particular form that is chosen for tractability in order to make the solution of the pricing equations (3.47) explicit. Contrary to this approach, we will take in the empirical part a more realistic expression for m_t even if pricing equations have to be solved numerically.

these changes in measure only shift the conditional Gaussian distribution of the log-returns thus the corresponding historical and risk-neutral conditional variances are equal in contradiction with classical empirical observations. Based on this assessment, Monfort and Pegoraro (2012) (see also Christoffersen et al. 2010, 2013 for analogous approaches) proposed an extension of the classical Esscher transform including a quadratic term in the pricing kernel to modify not only the conditional mean but also the conditional variance.[20] For Gaussian (or mixture of Gaussian) innovations, they also proved that the risk-neutral dynamics is explicit and different from the one implied by the LRNVR.

We assume for the SDF an exponential quadratic function of the log-returns: $\forall t \in \{1, \ldots, T\}$,

$$M_t = e^{\xi_t + \theta_{1,t} Y_t + \theta_{2,t} Y_t^2} \tag{3.66}$$

where $Y_t = log\left(\frac{S_t}{S_{t-1}}\right)$ and where ξ_t, $\theta_{1,t}$ and $\theta_{2,t}$ are $\mathcal{F}_{t-1}$ measurable random variables. For all $t \in \{1, \ldots, T\}$, we suppose that the conditional moment generating function, under $\mathbb{P}$, of (Y_t, Y_t^2) given $\mathcal{F}_{t-1}$ exists on $\mathbb{R}^2$ and is denoted by

$$\mathbb{G}^{\mathbb{P}}_{(Y_t, Y_t^2)|\mathcal{F}_{t-1}}(u, v) = E_{\mathbb{P}}\left[e^{uY_t + vY_t^2}\right].$$

In this case, the pricing equations (3.14) and (3.15) are written as

$$\xi_t = -r - log\left(\mathbb{G}^{\mathbb{P}}_{(Y_t, Y_t^2)|\mathcal{F}_{t-1}}(\theta_{1,t}, \theta_{2,t})\right) \tag{3.67}$$

$$\frac{\mathbb{G}^{\mathbb{P}}_{(Y_t, Y_t^2)|\mathcal{F}_{t-1}}(\theta_{1,t} + 1, \theta_{2,t})}{\mathbb{G}^{\mathbb{P}}_{(Y_t, Y_t^2)|\mathcal{F}_{t-1}}(\theta_{1,t}, \theta_{2,t})} = e^r. \tag{3.68}$$

Remark 3.5.1 We know from Proposition 3.4.1 that, in the case of the classical Esscher transform, the pricing equations have a unique solution. Here, this system has in general an infinite number of solutions (with the notations of the preceding section, the pair $(\theta_t^q, 0)$ fulfills Eq. (3.68)). In particular, from the implicit function theorem, $\theta_{2,t}$ is locally a regular function of $\theta_{1,t}$.[21]

---

[20]Even if it tends to contradict the monotonic pattern assumed in existing theory for the SDF, recent empirical studies show evidences on possible U-shapes speaking for quadratic parameterizations (Christoffersen et al. 2013).

[21]Let $f : \mathbb{R}^2 \to \mathbb{R}$ be a continuously differentiable function such that $f(x_0, y_0) = 0$ and $\frac{\partial f}{\partial y}(x_0, y_0) \neq 0$. The implicit function theorem ensures that there exists an open set U_{x_0} containing x_0, an open set V_{y_0} containing y_0 and a continuously differentiable function $g : U_{x_0} \to V_{y_0}$ such that

$$\{(x, g(x)) \mid x \in U_{x_0}\} = \{(x, y) \in U_{x_0} \times V_{y_0} \mid f(x, y) = 0\}.$$

From now on, we suppose $\forall t \in \{1, \ldots, T\}$ that Eq. (3.68) has a solution $(\theta_{1,t}^q, \theta_{2,t}^q)$ and we denote by $\mathbb{Q}^{qess}$ the EMM associated to the one-period SDF process (3.66). We obtain the following relation that describes the dynamics of the log-returns under $\mathbb{Q}^{qess}$:

$$\mathbb{G}_{Y_t|\mathcal{F}_{t-1}}^{\mathbb{Q}^{qess}}(u) = \frac{\mathbb{G}_{(Y_t, Y_t^2)|\mathcal{F}_{t-1}}^{\mathbb{P}}(u + \theta_{1,t}^q, \theta_{2,t}^q)}{\mathbb{G}_{(Y_t, Y_t^2)|\mathcal{F}_{t-1}}^{\mathbb{P}}(\theta_{1,t}^q, \theta_{2,t}^q)}. \tag{3.69}$$

In particular in the Gaussian case we obtain the following proposition:

Proposition 3.5.1 (Monfort and Pegoraro 2012) *If we assume that, in Eq. (3.4), the innovations are Gaussian and if* $\forall t \in \{1, \ldots, T\}$, $\theta_{2,t}^q < \frac{1}{2h_t}$,

(1) The functional relation between $\theta_{1,t}^q$ *and* $\theta_{2,t}^q$ *is global (see Remark 3.5.1) and explicit:*

$$\frac{h_t}{2(1 - 2\theta_{2,t}^q h_t)} + \frac{h_t \theta_{1,t}^q + r + m_t}{1 - 2\theta_{2,t}^q h_t} = r. \tag{3.70}$$

(2) Under $\mathbb{Q}^{qess}$,

$$\begin{cases} Y_t = r - \frac{h_t^*}{2} + \sqrt{h_t^*}\xi_t, \quad S_0 = s, \\ \frac{h_t^*}{1+2\theta_{2,t}^q h_t^*} = F\left(\sqrt{1 + 2\theta_{2,t-1}^q h_{t-1}^*}\left(-\frac{m_{t-1}}{\sqrt{h_{t-1}^*}} - \frac{\sqrt{h_{t-1}^*}}{2} + \xi_{t-1}\right), \frac{h_{t-1}^*}{1+2\theta_{2,t-1}^q h_{t-1}^*}\right) \end{cases} \tag{3.71}$$

where $h_t^* = \frac{h_t}{1-2\theta_{2,t}^q h_t}$ *and where, under* $\mathbb{Q}^{qess}$, *the* ξ_t *are i.i.d and* $\mathcal{F}_t$ *measurable random variables such that* ξ_t *follows a* $\mathcal{N}(0, 1)$.

Remark 3.5.2 If $\forall t \in \{1, \ldots, T\}$, $\theta_{2,t-1}^q = 0$, we recover the risk-neutral dynamics given by Proposition 3.2.2, in particular, $h_t = h_t^*$. Furthermore, if $\forall t \in \{1, \ldots, T\}$, $\theta_{2,t-1}^q > 0$ (implying a U-shape for (3.66)) then $h_t < h_t^*$: starting with a GARCH model with Gaussian innovations, the U-shaped quadratic Esscher transform induces a risk-neutral variance strictly greater than the historical one.

In our setting, we just take

$$f(x, y) = \mathbb{G}_{(Y_t, Y_t^2)|\mathcal{F}_{t-1}}^{\mathbb{P}}(x + 1, y) - e^r \mathbb{G}_{(Y_t, Y_t^2)|\mathcal{F}_{t-1}}^{\mathbb{P}}(x, y).$$

Example 3.5.1 Considering the Heston and Nandi (2000) model (3.38) and assuming a constant proportional wedge between h_t and h_t^* (i.e. $\frac{h_t^*}{h_t} = \pi > 0$) we have

$$1 + 2\theta_{2,t}^q h_t^* = \pi \text{ and } 1 - 2\theta_{2,t}^q h_t = \frac{1}{\pi}.$$

Thus, we obtain under $\mathbb{Q}^{qess}$,

$$\begin{cases} Y_t = r - \frac{h_t^*}{2} + \sqrt{h_t^*}\xi_t, \quad S_0 = s \\ h_t^* = \pi a_0 + \pi^2 a_1(\xi_{t-1} - (\frac{\gamma}{\pi} + \frac{\lambda_0}{\pi} + \frac{1}{2})\sqrt{h_{t-1}^*})^2 + b_1 h_{t-1}^* \end{cases} \tag{3.72}$$

where the ξ_t are i.i.d $\mathcal{N}(0,1)$ under $\mathbb{Q}^{qess}$. In this case, we recover the discrete time dynamics of Christoffersen et al. (2013). Moreover, since

$$cov_{\mathbb{Q}^{qess}}(h_{t+1}^*, Y_t \mid \mathcal{F}_{t-1}) = -2\pi^2 a_1(\gamma + \lambda_0 + \frac{\pi}{2})h_t$$

and

$$cov_{\mathbb{Q}^{ess}}(h_{t+1}, Y_t \mid \mathcal{F}_{t-1}) = -2a_1(\gamma + \lambda_0)h_t,$$

this new risk neutralization induces not only a bigger variance but also a greater leverage effect than the Esscher transform.[22]

The proof of Proposition 3.5.1 is based on the following lemma

Lemma 3.5.1 *If Y follows a $\mathcal{N}(\mu, \sigma^2)$, $\forall(u,v) \in \mathbb{R}^2$ such that $v < \frac{1}{2\sigma^2}$,*

$$\mathbb{E}(e^{uY+vY^2}) = e^{\frac{\sigma^2 u^2 + 2\mu u + 2v\mu^2}{2(1-2v\sigma^2)} - \frac{1}{2}log(1-2v\sigma^2)}. \tag{3.73}$$

The preceding result is due to the very particular form of the Gaussian density, that is why it seems to be hard to obtain the analogous of Proposition 3.5.1 in great generality.[23] Thus, working in the flexible quadratic Esscher transform setting is at the cost of very particular distributions for the innovations. In the empirical part, we test if this restriction improve pricing performances.

In the next section, we present a simple alternative to the preceding parametric specifications of the SDF that is based on the so-called empirical martingale simulation method.

---

[22]Even if it is not possible to estimate the free parameter π only using the log-returns (π not being a parameter of the physical dynamics) we will see in Sect. 3.7 that π may be efficiently extracted from option prices or in Chap. 4 that it may be estimated from the so-called VIX.

[23]Nevertheless, Monfort and Pegoraro (2012) proved that a natural extension is possible in the case of the mixture of Gaussian distributions.

3.6 The Empirical Martingale Simulation Method

The empirical martingale simulation (EMS) is an interesting variance reduction technique introduced in Duan and Simonato (1998) and widely used to improve the numerical efficiency of Monte Carlo estimators in GARCH option pricing models. It is based on the observation that once Monte Carlo methods are used to approximate expectations under one EMM, the associated empirical prices often violate no arbitrage bounds. A slight modification of the classical numerical scheme allows to overcome this problem.

Let us use the notations of Sect. 3.2. Considering an arbitrary EMM Q, the martingale property requires that $\forall t \in \{0, \ldots, T\}$,

$$E_Q[e^{-r(T-t)} S_T | \mathcal{F}_t] = S_t \tag{3.74}$$

and the arbitrage-free price process $(C_t(f))_{t\in\{0,\ldots,T\}}$ associated to an European contingent claim with payoff $f(S_T)$ is given by

$$E_Q[e^{-r(T-t)} f(S_T) | \mathcal{F}_t] = C_t(f). \tag{3.75}$$

In general (3.75) does not have a closed-form expression and Monte Carlo estimators (see the Mathematical Appendix) have to be used: if we denote by $(S_{T,i})_{i\in\{1,\ldots,n\}}$ a n-sample of the conditional distribution of S_T given $\mathcal{F}_t$ under Q, the Monte Carlo price estimate is written as

$$C_{t,n}^{MC}(f) = \frac{e^{-r(T-t)}}{n} \sum_{i=1}^{n} f(S_{T,i}). \tag{3.76}$$

As remarked in Duan and Simonato (1998), classical no arbitrage bounds are often violated once this approximation is used because $C_{t,n}^{MC}(f) \neq C_t(f)$, in particular, the martingale condition (3.74) is not empirically verified

$$E_Q[e^{-r(T-t)} S_T | \mathcal{F}_t] \approx C_{t,n}^{MC}(Id) \neq S_t. \tag{3.77}$$

To overcome the preceding problem and improve simulation accuracy, Duan and Simonato (1998) have proposed the EMS method that is a powerful and simple adjustment to Monte Carlo simulations that is equivalent to enforcing call-put parity.[24] They replace the i-*th* sampled final value $S_{T,i}$ by

[24] In fact, from (3.78), we deduce the empirical call-put parity

$$\frac{e^{-r(T-t)}}{n} \sum_{i=1}^{n} (\tilde{S}_{T,i} - K)_+ + S_t = Ke^{-r(T-t)} + \frac{e^{-r(T-t)}}{n} \sum_{i=1}^{n} (K - \tilde{S}_{T,i})_+$$

that is equivalent to (3.80).

$$\tilde{S}_{T,i} = \frac{S_{T,i}}{\frac{1}{n}\sum_{i=1}^{n} S_{T,i}} S_t e^{r(T-t)} \tag{3.78}$$

and the new price estimate becomes

$$C_{t,n}^{EMS}(f) = \frac{e^{-r(T-t)}}{n} \sum_{i=1}^{n} f(\tilde{S}_{T,i}). \tag{3.79}$$

In particular, the empirical martingale relation

$$C_{t,n}^{EMS}(Id) = S_t \tag{3.80}$$

is fulfilled. The consistency and the asymptotic normality of this estimator have been proved in Duan and Simonato (1998) and Duan et al. (2001) for Lipschitz payoffs and extended in Yuan and Chen (2009a,b) to a class of payoff functions which are continuous and with first derivatives bounded by some polynomial functions.[25] What is more, EMS is known to considerably reduce the empirical standard deviation of the price estimators (see Table 3.1 for a numerical example in the Black and Scholes setting) explaining why it has been employed in various studies especially in the GARCH framework (see, among others, Barone-Adesi et al. 2008; Chorro et al. 2012 and Guégan et al. 2013).

In Chorro et al. (2010), starting from the fact that the algebraic relation (3.80) remains true if the n-sample $(S_{T,i})_{i\in\{1,\ldots,n\}}$ is generated under $\mathbb{P}$ instead of under the EMM Q, the authors proposed a new path to risk neutralization and directly priced European options from the historical distribution of the log-returns and the

Table 3.1 Standard deviations for Monte Carlo $C_{0,1000}^{MC}$ and EMS $C_{0,1000}^{EMS}$ estimators of European call options of maturity T and strike K in the Black and Scholes model with parameters $r = 0.1$ (annualized), $\sigma = 0.2$ (annualized) and $S_0 = 100$

	T = 1 month			T = 6 months			T = 1 year		
$\frac{S_0}{K}$	0.9	1	1.1	0.9	1	1.1	0.9	1	1.1
MC	0.024	0.110	0.174	0.219	0.351	0.401	0.402	0.515	0.573
EMS	0.022	0.054	0.016	0.134	0.129	0.079	0.194	0.157	0.111

The standard deviations have been computed using 1,000 independent realizations

[25] Using the fact that prices may equivalently be expressed using expectations under the historical probability involving the stochastic discount factor (see (3.12)), Huang (2014) and Huang and Tu (2014) proposed and studied an historical version of the EMS when a risk-neutral model is not conveniently obtained. The price to pay is a computational cost that may be heavy if extra techniques are not used.

transform (3.78).[26] In their framework, the price at time t of a European call option with strike K and maturity T is given by

$$e^{-r(T-t)}\frac{1}{n}\sum_{i=1}^{n}(\tilde{S}_{T,i}-K)_+$$

where $(S_{T,i})_{i\in\{1,\dots,n\}}$ is a n-sample of the conditional distribution of S_T given $\mathcal{F}_t$ under $\mathbb{P}$. The promising empirical pricing performances of this new methodology have been studied in Chorro et al. (2010) and will be confirmed in Chap. 4.

The next subsection reviews closed-form pricing formulas that can be obtained in the GARCH setting in the spirit of Heston and Nandi (2000).

3.7 Remarks on Closed-Form Option Pricing Formulas

It is well-known that in the Black and Scholes (1973) model, the price of European Call and Put options is an explicit function of the volatility parameter. Nevertheless, in most of realistic cases, Monte Carlo estimators are needed in order to compute option prices as in Duan (1995). In a seminal paper, Heston and Nandi (2000) studied a particular GARCH model with Gaussian innovations that replicates one of key features of competing continuous time models: the fact that the characteristic function of the log-returns under the risk neutral distribution has a closed-form expression.[27] Thus, in their model, options are priced efficiently from the powerful fast Fourier transform (FFT) (see Walker 1996) as explained in Carr and Madan (1999). In particular, the parameters of the model may be calibrated from option datasets at a reasonable computational cost. The key stone is to consider a GARCH in mean model such that both conditional expectation and variance of the volatility process are affine functions of the volatility at the preceding trading date (we speak about affine models[28]). We review in this section the rough outlines of this approach.[29]

---

[26] Here, there is no particular assumption on the shape of the SDF but relation (3.78) may be seen as an empirical pricing equation.

[27] To be more precise, Heston and Nandi (2000) proved an explicit backward-recursive equation to compute efficiently the moment generating function of the log-returns (see Proposition 3.7.1).

[28] See for instance Hsieh and Ritchken (2005) for comparisons of linear and non-linear modelling in the Gaussian case.

[29] In Christoffersen et al. (2006) and Mercuri (2008), the authors use the same idea with some non-Gaussian innovations combined with very particular affine GARCH specifications.

If $\mathbb{Q}$ is an EMM we see, following Heston and Nandi (2000) (see also the Mathematical Appendix), that the price at time t of a call option with strike K and maturity T is given by

$$e^{-r(T-t)}E_{\mathbb{Q}}[(S_T-K)_+ \mid \mathcal{F}_t] = \frac{S_t}{2} + \frac{e^{-r(T-t)}}{\pi}\int_0^{+\infty} \mathcal{R}e\left[\frac{K^{-i\phi}\mathbb{G}^{\mathbb{Q}}_{log(S_T)|\mathcal{F}_t}(i\phi+1)}{i\phi}\right]d\phi$$
$$- Ke^{-r(T-t)}\left(\frac{1}{2} + \frac{1}{\pi}\int_0^{+\infty} \mathcal{R}e\left[\frac{K^{-i\phi}\mathbb{G}^{\mathbb{Q}}_{log(S_T)|\mathcal{F}_t}(i\phi)}{i\phi}\right]d\phi\right).$$

Up to numerical integration, we have to evaluate $\mathbb{G}^{\mathbb{Q}}_{log(S_T)|\mathcal{F}_t}$ efficiently to compute this price. Let us see an example.

For the Heston and Nandi (2000) model (3.38), we have seen that the risk-neutral dynamics under the equilibrium EMM is given by Proposition 3.2.2:

$$\begin{cases} Y_t = r - \frac{h_t}{2} + \sqrt{h_t}\xi_t, \quad S_0 = s \\ h_t = a_0 + a_1(\xi_{t-1} - \underbrace{(\gamma + \lambda_0 + \frac{1}{2})}_{\gamma^*}\sqrt{h_{t-1}})^2 + b_1 h_{t-1} \end{cases} \tag{3.81}$$

where the ξ_t are i.i.d $\mathcal{N}(0,1)$. Then, we have the following proposition where $\mathbb{G}^{\mathbb{Q}}_{log(S_T)|\mathcal{F}_t}$ is obtained from difference equations:

Proposition 3.7.1 *Under the dynamics* (3.81),

$$\mathbb{G}^{\mathbb{Q}}_{log(S_T)|\mathcal{F}_t}(u) = S_t^u e^{A_t + B_t h_{t+1}}$$

where

$$A_t = ru + A_{t+1} + a_0 B_{t+1} - \frac{1}{2}log(1 - 2a_1 B_{t+1})$$

and

$$B_t = \frac{-1}{2}u + b_1 B_{t+1} + \frac{\frac{u^2}{2} - 2a_1\gamma^* B_{t+1}u + a_1 B_{t+1}(\gamma^*)^2}{1 - 2a_1 B_{t+1}}$$

with terminal conditions $A_T = B_T = 0$.

Remark 3.7.1 In the framework of Example 3.5.1, we obtain the analogous of the preceding proposition replacing $(a_0, a_1, \gamma^*, b_1)$ by $(\pi a_0, \pi^2 a_1, \frac{\gamma}{\pi} + \frac{\lambda_0}{\pi} + \frac{1}{2}, b_1)$. In particular, the proportional wedge between the historical and the risk-neutral volatilities may be efficiently estimated from option prices.

Analogous results may be obtained in the same spirit for the inverse Gaussian and the tempered stable innovations. We refer the reader to Christoffersen et al. (2006) and Mercuri (2008) for details.

3.8 Proofs of Chapter 3

Proof of Proposition 3.2.2 From Definition 3.2.4, Y_t may be written as

$$Y_t = \nu_t + \sqrt{h_t}\xi_t$$

where $\nu_t = E_Q[Y_t \mid \mathcal{F}_{t-1}]$ and where, given $\mathcal{F}_{t-1}$, ξ_t follows a $\mathcal{N}(0,1)$. We deduce from the martingale condition $E_Q[e^{Y_t} \mid \mathcal{F}_{t-1}] = e^r$ that $\nu_t = r - \frac{h_t}{2}$. Using Eq. (3.4), we obtain that

$$\xi_{t-1} = \frac{r + m_{t-1} - \nu_{t-1}}{\sqrt{h_{t-1}}} + z_{t-1}$$

and the dynamics of the conditional volatility may be explicitly expressed as a function of ξ_{t-1} as in Eq. (3.36). □

Proof of Lemma 3.3.1 Let X be a real random variable defined on Ω that is $\sigma(e^Z)$ measurable. Thus, $X = h(e^Z)$ where $h : \mathbb{R} \to \mathbb{R}$ is Borelian. We have

$$E_{\mathbb{Q}_m}[X] = \int_{\mathbb{R}} h(x)\tilde{d}(x)dx = \int_{\mathbb{R}} h(x)\frac{\tilde{d}(x)}{d(x)}d(x)dx = E_{\mathbb{P}}\left[X\frac{\tilde{d}(e^Z)}{d(e^Z)}\right]$$

and

$$\frac{d\mathbb{Q}_m}{d\mathbb{P}} = \frac{\tilde{d}(e^Z)}{d(e^Z)}.$$

The conclusion follows remarking that $\tilde{d}(x) = d(xe^m)e^m$. Moreover, supposing that $m = log(E_{\mathbb{P}}[e^Z])$, we deduce from

$$E_{\mathbb{Q}_m}[e^Z] = E_{\mathbb{P}}[e^{Z-m}]$$

that $E_{\mathbb{Q}_m}[e^Z] = 1$. □

Proof of Proposition 3.4.1 To obtain the uniqueness of the solution, it is sufficient to prove that the mapping Φ_t is strictly increasing. Computing the derivative we obtain

$$\Phi_t'(\theta) = g_t(\theta + 1) - g_t(\theta)$$

where

$$g_t(\theta) = \frac{\left(\mathbb{G}^{\mathbb{P}}_{Y_{t+1}|\mathcal{F}_t}\right)'(\theta)}{\mathbb{G}^{\mathbb{P}}_{Y_{t+1}|\mathcal{F}_t}(\theta)}.$$

We can see, using the Proposition A.4 of the Appendix with the nonnegative martingale

$$\left(\prod_{k=1}^{t} \frac{e^{\theta Y_k}}{\mathbb{G}^{\mathbb{P}}_{Y_k|\mathcal{F}_{k-1}}(\theta)}\right)_{t\in\{1,\ldots,T\}},$$

that

$$(g_t)'(\theta) = Var_{\tilde{\mathbb{P}}}[Y_{t+1} \mid \mathcal{F}_t] > 0$$

where

$$\frac{d\tilde{\mathbb{P}}}{d\mathbb{P}} = \prod_{k=1}^{T} \frac{e^{\theta Y_k}}{\mathbb{G}^{\mathbb{P}}_{Y_k|\mathcal{F}_{k-1}}(\theta)}.$$

Thus, g_t is strictly increasing and $\Phi' > 0$. The conclusion follows. □

Proof of Proposition 3.4.4 (1) From the Kolmogorov representation theorem (see Mainardi and Rogosin 2006 for an interesting historical approach to this result), we have for all $u \in \mathcal{D}_{\mathbb{G}^{\mathbb{P}}_{Y_t|\mathcal{F}_{t-1}}}$,

$$log(\mathbb{G}^{\mathbb{P}}_{Y_t|\mathcal{F}_{t-1}}(u)) = \gamma_t u + \int_{-\infty}^{+\infty} (e^{zu} - 1 - zu)\frac{dK_t(z)}{z^2}$$

where γ_t is a $\mathcal{F}_{t-1}$ measurable real random variable and K_t a $\mathcal{F}_{t-1}$ measurable random variable with values in the space of the non-decreasing and bounded functions with limit zero in $-\infty$. Thus, from Proposition 3.4.2, $\forall u \in \{\theta \in \mathbb{R};\ \theta + \theta_t^q \in \mathcal{D}_{\mathbb{G}^{\mathbb{P}}_{Y_t|\mathcal{F}_{t-1}}}\}$,

$$log(E_{\mathbb{Q}^{ess}}[e^{uY_t} \mid \mathcal{F}_{t-1}]) = \gamma_t u + \int_{-\infty}^{+\infty} (e^{z(u+\theta_t^q)} - e^{\theta_t^q z} - zu)\frac{dK_t(z)}{z^2}$$

thus

$$log(E_{\mathbb{Q}^{ess}}[e^{uY_t} \mid \mathcal{F}_{t-1}]) = \tilde{\gamma}_t u + \int_{-\infty}^{+\infty} (e^{zu} - 1 - uz)\frac{e^{\theta_t^q z} dK_t(z)}{z^2} \tag{3.82}$$

where

$$\tilde{\gamma}_t = \gamma_t + \int_{-\infty}^{+\infty} (e^{\theta_t^q z} - 1) z \frac{dK_t(z)}{z^2}.$$

Since $\mathbb{G}^{\mathbb{P}}_{Y_t|\mathcal{F}_{t-1}}$ is twice differentiable, we have in particular that

$$\int_{-\infty}^{+\infty} e^{\theta_t^q z} dK_t(z) < \infty$$

and we may define $\forall x \in \mathbb{R}$,

$$\tilde{K}_t(x) = \int_{-\infty}^{x} e^{\theta_t^q z} dK_t(z)$$

that is a $\mathcal{F}_{t-1}$ measurable random variable with values in the space of the non-decreasing and bounded functions with limit zero in $-\infty$. The conclusion follows from (3.82) and from the Kolmogorov representation theorem.

(2) From the hypothesis, the conditional density, under $\mathbb{P}$, of z_t given $\mathcal{F}_{t-1}$ is of the form

$$f^{\mathbb{P}}_{z_t|\mathcal{F}_{t-1}}(x) = \sum_{i=1}^{N} a_i f_i(x)$$

where the a_i are nonnegative and $\mathcal{F}_{t-1}$ measurable real random variables such that $\sum_{i=1}^{N} a_i = 1$ and where the f_i are probability densities on $\mathbb{R}$. Let us define $\forall i \in \{1, \ldots, N\}$, a $\mathcal{F}_t$ measurable random variable Z_t^i having, under $\mathbb{P}$, the conditional density f_i given $\mathcal{F}_{t-1}$. Thus,

$$E_{\mathbb{P}}[e^{uz_t} \mid \mathcal{F}_{t-1}] = \sum_{i=1}^{N} a_i E_{\mathbb{P}}[e^{uZ_t^i} \mid \mathcal{F}_{t-1}].$$

From Proposition 3.4.2,

$$E_{\mathbb{Q}^{ess}}[e^{uz_t} \mid \mathcal{F}_{t-1}] = \frac{E_{\mathbb{P}}[e^{(u+\sqrt{h_t}\theta_t^q)z_t} \mid \mathcal{F}_{t-1}]}{E_{\mathbb{P}}[e^{\sqrt{h_t}\theta_t^q z_t} \mid \mathcal{F}_{t-1}]}$$

so

$$E_{\mathbb{Q}^{ess}}[e^{uz_t} \mid \mathcal{F}_{t-1}] = \sum_{i=1}^{N} a_{i,\tilde{t}-1} E_{\mathbb{P}}\left[e^{uZ_t^i} \frac{e^{\sqrt{h_t}\theta_t^q Z_t^i}}{E_{\mathbb{P}}[e^{\sqrt{h_t}\theta_t^q Z_t^i} \mid \mathcal{F}_{t-1}]} \mid \mathcal{F}_{t-1} \right]$$

where

$$a_{i,\tilde{t}-1} = \frac{a_i E_{\mathbb{P}}[e^{\sqrt{h_t}\theta_t^q Z_t^i} \mid \mathcal{F}_{t-1}]}{\sum_{i=1}^{N} a_i E_{\mathbb{P}}[e^{\sqrt{h_t}\theta_t^q Z_t^i} \mid \mathcal{F}_{t-1}]}$$

is $\mathcal{F}_{t-1}$ measurable. Finally, we can write

$$f_{z_t|\mathcal{F}_{t-1}}^{\mathbb{Q}^{ess}}(x) = \sum_{i=1}^{N} a_{i,\tilde{t}-1} f_{i,\tilde{t}-1}^{\mathbb{Q}^{ess}}(x)$$

where $f_{i,t-1}^{\mathbb{Q}^{ess}}$ is the conditional Esscher transform of parameter $\sqrt{h_t}\theta_t^q$ of the conditional density of Z_t^i under $\mathbb{P}$. This gives the conclusion. □

Proof of Proposition 3.4.5 For $\mid \beta + u \mid < \alpha$,

$$\mathbb{G}_{GH}(u) = e^{\mu u}\left(\frac{\alpha^2 - \beta^2}{\alpha^2 - (\beta + u)^2}\right)^{\frac{\lambda}{2}} \frac{K_\lambda(\delta\sqrt{\alpha^2 - (\beta + u)^2})}{K_\lambda(\delta\sqrt{\alpha^2 - \beta^2})},$$

hence $\mathbb{G}_{GH}$ is twice differentiable. Moreover, $\Phi(\theta) = log\left(\frac{\mathbb{G}_{GH}(1+\theta)}{\mathbb{G}_{GH}(\theta)}\right)$ is defined on the interval $]-(\alpha+\beta), \alpha - \beta - 1[$ that is not empty because $\alpha > \frac{1}{2}$. Thus we may apply Proposition 3.4.1 and the uniqueness holds. The existence still needs to be proved.

(1) For $x > 0$, we define $\Psi(x) = log\left(\frac{K_\lambda(x)}{x^\lambda}\right)$. Thus,

$$\Phi(\theta) = \mu + \Psi(\delta\sqrt{\alpha^2 - (\beta + 1 + \theta)^2}) - \Psi(\delta\sqrt{\alpha^2 - (\beta + \theta)^2}).$$

For the properties of the Bessel function used in the sequel we refer the reader to Abramowitz and Stegun (1964). If $\lambda > 0$,

$$\frac{K_\lambda(x)}{x^\lambda} \sim_{x\to 0+} \frac{\Gamma[\lambda]2^{\lambda-1}}{x^{2\lambda}}.$$

So we have $\lim_{\theta\to\alpha-\beta-1} \Phi(\theta) = +\infty$ and $\lim_{\theta\to-(\alpha+\beta)} \Phi(\theta) = -\infty$. The conclusion follows from the intermediate value theorem. When $\lambda = 0$, we may conclude as before, remembering that $K_0(x) \sim_{x\to 0+} -\log(x/2) - \gamma$ where γ is the Euler–Mascheroni constant.

(2) When $\lambda < 0$, using the relation $K_\lambda(x) = K_{-\lambda}(x)$ we obtain that

$$\frac{K_\lambda(x)}{x^\lambda} \sim_{x\to 0+} \Gamma[-\lambda]2^{-\lambda-1}.$$

Thus, $\lim_{\theta\to\alpha-\beta-1} \Phi(\theta) = \mu + C$ and $\lim_{\theta\to-(\alpha+\beta)} \Phi(\theta) = \mu - C$ and we conclude applying again the intermediate value theorem. □

Proof of Proposition 3.4.7 The point (3) is a direct consequence of the expression of $\mathbb{G}_{GH}$ for $\lambda = -\frac{1}{2}$ that may be found in Eberlein and Hammerstein (2003).

Let us do the proof of point (2), point (1) being obtained in the same way. As mentioned in Remark 3.4.3 we use for simplicity the unconstrained specification of the model. Thus, Eq. (3.4) becomes

$$Y_t = log\left(\frac{S_t}{S_{t-1}}\right) = r + m_t + \sqrt{h_t}\left(\frac{z_t - E_{\mathbb{P}}[z_t]}{\sqrt{Var_{\mathbb{P}}[z_t]}}\right), \quad S_0 = s \in \mathbb{R}_+,$$

where z_t follows a $GH(\lambda, \alpha, \beta, \delta, \mu)$ and

$$\begin{cases} E_{\mathbb{P}}[z_t] = \mu + \frac{\delta\beta}{\gamma}\frac{K_{\lambda+1}(\delta\gamma)}{K_\lambda(\delta\gamma)} \\ Var_{\mathbb{P}}[z_t] = \frac{\delta}{\gamma}\frac{K_{\lambda+1}(\delta\gamma)}{K_\lambda(\delta\gamma)} + \frac{\beta^2\delta^2}{\gamma^2}\left(\frac{K_{\lambda+2}(\delta\gamma)K_\lambda(\delta\gamma) - K^2_{\lambda+1}(\delta\gamma)}{K^2_\lambda(\delta\gamma)}\right) \end{cases}$$

with $\gamma = \sqrt{\alpha^2 - \beta^2}$. Remarking that

$$Y_t = r + \tilde{m}_t + \sqrt{\tilde{h}_t} z_t$$

where

$$\begin{cases} \tilde{h}_t = \frac{h_t}{Var_{\mathbb{P}}[z_t]} \\ \tilde{m}_t = m_t - \sqrt{\tilde{h}_t}\, E_{\mathbb{P}}[z_t] \end{cases}$$

we deduce from Eqs. (3.55) and (3.56) that under $\mathbb{Q}^{ess}$,

$$Y_t = r + \tilde{m}_t + \sqrt{\tilde{h}_t}\xi_t, \quad S_0 = s,$$

where the ξ_t are $\mathcal{F}_t$ measurable random variables such that, conditionally to $\mathcal{F}_{t-1}$,

$$\xi_t \hookrightarrow GH(\lambda, \alpha, \beta + \sqrt{\tilde{h}_t}\theta^q_t, \delta, \mu).$$

When $\mu = 0, \lambda = a > 0, \alpha, \beta \to +\infty, \alpha\delta^2 \to 0$ and $\alpha - \beta = b > 0$, we deduce from Eberlein and Hammerstein (2003) that

$$\begin{cases} \tilde{h}_t \to \frac{b^2 h_t}{a} \\ \tilde{m}_t \to m_t - \sqrt{a h_t} \\ \theta_t^q \to \sqrt{\frac{a}{h_t}} - \left(1 - \frac{e^{m_t} - \sqrt{a h_t}}{a}\right)^{-1}. \end{cases}$$

From Lemma 3.4.1, under $\mathbb{Q}^{ess}$, the conditional distribution of ξ_t given $\mathcal{F}_{t-1}$ converges in distribution toward a

$$G_a\left(a, \sqrt{\frac{b^2 h_t}{a}}\left(1 - \frac{e^{m_t} - \sqrt{a h_t}}{a}\right)^{-1}\right).$$

At the limit, we recover the risk-neutral dynamics of Siu et al. (2004). □

Proof of Proposition 3.4.8 In the proof we will denote $p_1 = \phi$ and $p_2 = 1 - \phi$.

(1) The equation $log\left(\frac{\mathbb{G}_{MN}(1+\theta)}{\mathbb{G}_{MN}(\theta)}\right) = r$ is equivalent to

$$K(\theta) = \sum_{i=1}^{2} p_i e^{\theta\mu_i + \frac{\sigma_i^2\theta^2}{2}}\left[e^{\mu_i + \frac{\sigma_i^2}{2} + \sigma_i^2\theta} - e^r\right] = 0.$$

The distribution being centered, μ_1 or μ_2 is strictly negative and one obtains $K(\theta) \underset{\theta\to+\infty}{\to +\infty}$ and $K(\theta) \underset{\theta\to-\infty}{\to -\infty}$. The existence is a consequence of the intermediate value theorem and the uniqueness of Proposition 3.4.1.

(2) We just have to remark that

$$\frac{\mathbb{G}^{\mathbb{P}}_{Y_t|\mathcal{F}_{t-1}}(\theta_t^q + u)}{\mathbb{G}^{\mathbb{P}}_{Y_t|\mathcal{F}_{t-1}}(\theta_t^q)} = \sum_{i=1}^{2} \frac{p_i e^{\theta_t^q \mu_{i,t} + h_t\sigma_i^2\frac{(\theta_t^q)^2}{2}}}{\phi e^{\theta_t^q \mu_{1,t} + h_t\sigma_1^2\frac{(\theta_t^q)^2}{2}} + (1-\phi)e^{\theta_t^q \mu_{2,t} + h_t\sigma_2^2\frac{(\theta_t^q)^2}{2}}} e^{u(\mu_{i,t} + h_t\sigma_i^2\theta_t^q) + \frac{u^2 h_t\sigma_i^2}{2}}$$

where $\mu_{i,t} = r + m_t + \sqrt{h_t}\mu_i$. The result follows from Proposition 3.4.2. □

Proof of Proposition 3.4.9 According to the hypotheses,

$$\mathbb{G}^{\mathbb{P}}_{Y_t|\mathcal{F}_{t-1}}(u) = e^{\left[(r+m_t)u + \frac{u^2 h_t}{2} + \lambda\left(e^{\mu\sqrt{h_t}u + \frac{u^2 h_t \sigma^2}{2}} - 1\right)\right]}$$

thus

$$\frac{\mathbb{G}^{\mathbb{P}}_{Y_t|\mathcal{F}_{t-1}}(u+\theta^q_t)}{\mathbb{G}^{\mathbb{P}}_{Y_t|\mathcal{F}_{t-1}}(\theta^q_t)} = e^{\left[(r+m_t+\theta^q_t h_t)u+\frac{u^2 h_t}{2}+\lambda e^{\mu\theta^q_t\sqrt{h_t}+\frac{(\theta^q_t)^2 h_t\sigma^2}{2}}\left(e^{(\mu\sqrt{h_t}+\theta^q_t h_t\sigma^2)u+\frac{u^2 h_t\sigma^2}{2}}-1\right)\right]}$$

and the result follows from Proposition 3.4.2. □

Proof of Lemma 3.5.1 We write $Y=\mu+\sigma X$ where X follows a standard Gaussian distribution. Thus,

$$\begin{aligned}\mathbb{E}(e^{uY+vY^2}) &= e^{u\mu+v\mu^2}\mathbb{E}\left[e^{u\sigma X+v\sigma^2X^2+2v\mu\sigma X}\right]\\ &= e^{u\mu+v\mu^2-v\sigma^2\left(\frac{u+2v\mu}{2v\sigma}\right)^2}\mathbb{E}\left[e^{v\sigma^2\left(X+\frac{u+2v\mu}{2v\sigma}\right)^2}\right].\end{aligned}$$

Remembering that, when Z follows a $\mathcal{N}(0,1)$, we have for $a<\frac{1}{2}$ and $\forall b\in\mathbb{R}$

$$E[e^{a(Z+b)^2}] = e^{-\frac{1}{2}log(1-2a)+\frac{ab^2}{1-2a}},$$

then, for $v<\frac{1}{2\sigma^2}$

$$\begin{aligned}\mathbb{E}(e^{uY+vY^2}) &= e^{u\mu+v\mu^2-v\sigma^2\left(\frac{u+2v\mu}{2v\sigma}\right)^2}e^{-\frac{1}{2}log(1-2v\sigma^2)+\frac{v\sigma^2\left(\frac{u+2v\mu}{2v\sigma}\right)^2}{1-2v\sigma^2}}\\ &= e^{\frac{\sigma^2u^2+2\mu u+2v\mu^2}{2(1-2v\sigma^2)}-\frac{1}{2}log(1-2v\sigma^2)}.\end{aligned}$$

□

Proof of Proposition 3.5.1 Using Lemma 3.5.1, Eq. (3.68) is equivalent to

$$e^{\frac{h_t}{2(1-2\theta^q_{2,t}h_t)}+\frac{h_t\theta^q_{1,t}+r+m_t}{1-2\theta^q_{2,t}h_t}} = e^r$$

and Eq. (3.69) equivalent to

$$\mathbb{G}^{\mathbb{Q}^{qess}}_{Y_t|\mathcal{F}_{t-1}}(u) = e^{\frac{u^2h_t}{2(1-2\theta^q_{2,t}h_t)}+u\frac{h_t\theta^q_{1,t}+r+m_t}{1-2\theta^q_{2,t}h_t}}.$$

Thus, using Eq. (3.70),

$$\mathbb{G}^{\mathbb{Q}^{qess}}_{Y_t|\mathcal{F}_{t-1}}(u) = e^{\frac{u^2h_t}{2(1-2\theta^q_{2,t}h_t)}+u\left(r-\frac{h_t}{2(1-2\theta^q_{2,t}h_t)}\right)}.$$

Finally, we can write

$$Y_t = r - \frac{h_t^*}{2} + \sqrt{h_t^*}\xi_t, \quad S_0 = s,$$

where $h_t^* = \frac{h_t}{1-2\theta_{2,t}^q h_t}$ and where, under $\mathbb{Q}^{qess}$, the ξ_t are i.i.d and $\mathcal{F}_t$ measurable random variables such that ξ_t follows a $\mathcal{N}(0,1)$. The result follows from Eq. (3.4) remarking that

$$z_t = -\frac{2m_t + h_t^*}{2\sqrt{h_t}} + \sqrt{\frac{h_t^*}{h_t}}\xi_t.$$

□

Proof of Proposition 3.7.1 From $\mathbb{G}^{\mathbb{Q}}_{log(S_T)|\mathcal{F}_T}(u) = S_T^u$ we obtain the terminal conditions. Then, we do a backward induction reasoning. We have by the law of iterated expectations and the induction hypothesis

$$\mathbb{G}^{\mathbb{Q}}_{log(S_T)|\mathcal{F}_t}(u) = E_{\mathbb{Q}}[\mathbb{G}^{\mathbb{Q}}_{log(S_T)|\mathcal{F}_{t+1}}(u) \mid \mathcal{F}_t] = E_{\mathbb{Q}}[S_{t+1}^u e^{A_{t+1}+B_{t+1}h_{t+2}} \mid \mathcal{F}_t].$$

From (3.81),

$$\begin{aligned}
\mathbb{G}^{\mathbb{Q}}_{log(S_T)|\mathcal{F}_t}(u) &= S_t^u e^{ru+A_{t+1}+a_0 B_{t+1}+(\frac{-u}{2}+b_1 B_{t+1})h_{t+1}} \\
&\quad \times E_{\mathbb{Q}}[e^{u\sqrt{h_{t+1}}\xi_{t+1}+B_{t+1}a_1(\xi_{t+1}-\gamma^*\sqrt{h_{t+1}})^2} \mid \mathcal{F}_t] \\
&= S_t^u e^{ru+A_{t+1}+a_0 B_{t+1}+(\frac{-u}{2}+b_1 B_{t+1})h_{t+1}-a_1 B_{t+1}h_{t+1}\left((\frac{u}{2a_1 B_{t+1}}-\gamma^*)^2-(\gamma^*)^2\right)} \\
&\quad \times E_{\mathbb{Q}}[e^{B_{t+1}a_1(\xi_{t+1}+(\frac{u}{2a_1 B_{t+1}}-\gamma^*)\sqrt{h_{t+1}})^2} \mid \mathcal{F}_t].
\end{aligned}$$

Remembering that, when Z follows a $\mathcal{N}(0,1)$,

$$E[e^{a(Z+b)^2}] = e^{-\frac{1}{2}log(1-2a)+\frac{ab^2}{1-2a}},$$

we have

$$\begin{aligned}
\mathbb{G}^{\mathbb{Q}}_{S_T|\mathcal{F}_t}(u) &= S_t^u e^{ru+A_{t+1}+a_0 B_{t+1}-\frac{1}{2}log(1-2a_1 B_{t+1})} \\
&\quad \times e^{h_{t+1}(\frac{-u}{2}+b_1 B_{t+1}+\frac{\frac{u^2}{2}+a_1 B_{t+1}(\gamma^*)^2-2u\gamma^* a_1 B_{t+1}}{1-2a_1 B_{t+1}})}
\end{aligned}$$

and the result follows by identification. □

References

Abramowitz M, Stegun I (1964). Handbook of mathematical functions with formulas, graphs and mathematical tables. Dover, New York

Aît-Sahalia Y, Lo A (1998) Nonparametric estimation of state-price densities implicit in financial asset prices. J Financ 53:499–547

Aît-Sahalia Y, Lo A (2000) Nonparametric risk management and implied risk aversion. J Econ 94:9–51

Alexander C, Lazar E (2006) Normal mixture GARCH(1,1): applications to exchange rate modelling. J Appl Econ 21:307–336

Back K, Pliska SR (1991) On the fundamental theorem of asset pricing with an infinite state space. J Math Econ 20(1):1–18

Badescu A, Kulperger R (2008) GARCH option pricing: a semiparametric approach. Insur Math Econ 43(1):69–84

Badescu A, Kulperger R, Lazar E (2008) Option valuation with normal mixture GARCH. Stud Nonlinear Dynam Econ 12(2):1–42

Badescu A, Elliott RJ, Siu TK (2009) Esscher transforms and consumption-based models. Insur Math Econ 45(3):337–347

Badescu A, Elliott RJ, Kulperger R, Miettinen J, Siu TK (2011) A comparison of pricing kernels for GARCH option pricing with generalized hyperbolic distributions. Int J Theor Appl Financ 14(5):669–708

Barndorff-Nielsen OE (1995) Normal inverse Gaussian processes and the modelling of stock returns. Research Report 300, Department Theoretical Statistics, Aarhus University

Barndorff-Nielsen OE, Blaesild P (1981) Hyperbolic distributions and ramifications: contributions to theory and applications. In: Taillie C, Patil GP, Baldessari BA (eds) Statistical distributions in scientific work, vol 4. Reidel, Dordrecht, pp 19–44

Barone-Adesi G, Engle RF, Mancini LA (2008) GARCH option pricing model in incomplete markets. Rev Financ Stud 21(3):1223–1258

Bertholon H, Monfort A, Pegoraro F (2008) Econometric asset pricing modelling. J Financ Econ 6(4):407–458

Black F (1975) Fact and fantasy in the use of options. Financ Anal J 31(4):36–41

Black F, Scholes M (1973) The pricing of options and corporate liabilities. J Polit Econ 81:637–659

Bouchaud JP, Potters M (2003) Theory of financial risk and derivative pricing: from statistical physics to risk management. Cambridge University Press, Cambridge

Breeden DT, Gibbons MR, Litzenberger RH (1989) Empirical test of the consumption-oriented CAPM. J Financ 44(2):231–262

Brennan MJ (1979) The pricing of contingent claims in discrete time models. J Financ 24(1):53–68

Bühlmann H, Delbaen F, Embrechts P, Shiryaev AN (1996) No-arbitrage change of measure and conditional Esscher transforms. CWI Q 9:291–317

Carr P, Madan D (1999) Option valuation using the fast fourier transform. J Comput Financ 2(4):61–73

Chorro C, Guégan D, Ielpo F (2010) Martingalized historical approach for option pricing. Financ Res Lett 7(1):24–28

Chorro C, Guégan D, Ielpo F (2012) Option pricing for GARCH type models with generalized hyperbolic innovations. Quant Financ 12(7):1079–1094

Christoffersen P, Heston SL, Jacobs K (2006) Option valuation with conditional skewness. J Econ 131:253–284

Christoffersen P, Jacobs K, Ornthanalai C, Wang Y (2008) Option valuation with long-run and short-run volatility components. J Financ Econ 90:272–297

Christoffersen P, Elkamhi R, Feunou B, Jacobs K (2010) Option valuation with conditional heteroskedasticity and non-normality. Rev Financ Stud 23(5):2139–2183

Christoffersen P, Jacobs K, Ornthanalai C (2012) Dynamic jump intensities and risk premiums: evidence from S&P500 returns and options. J Financ Econ 106(3):447–472

Christoffersen P, Heston SL, Jacobs K (2013) Capturing option anomalies with a variance-dependent pricing kernel. Rev Financ Stud 26(8):1963–2006
Cochrane J (2001) Asset pricing. Princeton University Press, Princeton
Dalang RC, Morton A, Willinger W (1990) Equivalent martingale measures and no-arbitrage in stochastic securities market models. Stochast Stochast Rep 29(2):185–201
Delbaen F, Schachermayer W (2006) The mathematics of arbitrage. Springer finance series. Springer, Berlin
Duan JC (1995) The GARCH option pricing model. Math Financ 5:13–32
Duan JC, Simonato JG (1998) Empirical martingale simulation for asset prices. Manag Sci 44:1218–1233
Duan JC, Gauthier G, Simonato JG (2001) Asymptotic distribution of the EMS option price estimator. Manag Sci 47:1122–1132
Duan JC, Ritchken P, Sun Z (2005) Jump starting GARCH: pricing and hedging options with jumps in returns and volatilities. Working Paper, University of Toronto
Duan JC, Ritchken P, Sun Z (2006) Approximating GARCH-jump models, jump diffusion processes, and option pricing. Math Financ 16:21–52
Eberlein E, Hammerstein EAV (2003) Generalized hyperbolic and inverse Gaussian distributions: limiting cases and approximation of processes. In: Proceedings of the 4th Ascona conference. Birkhauser, Boston
Eberlein E, Prause K (2002) The generalized hyperbolic model: financial derivatives and risk measures. In: Geman H, Madan D, Pliska S, Vorst T (eds) Mathematical finance-bachelier congress 2000. Springer, Heidelberg, pp 245–267
Elliott R, Madan D (1998) A discrete time equivalent martingale measure. Math Financ 2(8):127–152
Elliott R, Siu TK, Chan L (2006) Option pricing for GARCH models with Markov switching. Int J Theor Appl Financ 9(6):825–841
Embretchs P, Frey R, McNeil AJ (2005) Quantitative risk management. Princeton University Press, Princeton
Esscher F (1932) On the probability function in the collective theory of risk. Skandinavisk Aktuarietidskrift 15:175–195
Föllmer H, Schweiser M (1991) Hedging of contingent claims under incomplete information. In: Davis MHA, Elliott RJ (eds) Applied stochastic analysis. Gordon and Breach, London, pp 389–414
Gerber HU, Shiu SW (1994a) Option pricing by Esscher transforms. Trans Soc Actuar 46:99–191
Gerber HU, Shiu SW (1994b) Martingale approach to pricing perpetual American options. In: Proceedings of the 4th AFIR international colloqium, transaction of Society of Actuaries, Orlando, pp 659–689
Gouriéroux C, Monfort A (2007) Econometric specification of stochastic discount factor models. J Econ 136(2):509–530
Guégan D, Ielpo F, Lalaharison H (2013) Option pricing with discrete time jump processes. J Econ Dyn Control 37(12):2417–2445
Hansen LP, Richard SF (1987) The role of conditioning information in deducing testable restrictions implied by dynamic asset pricing models. Econometrica 55(3):587–613
Harrison JM, Kreps DM (1979) Martingales and arbitrage in multiperiod securities markets. J Econ Theory 20(3):381–408
Harrison JM, Pliska SR (1981) Martingales and stochastic integrals in the theory of continuous trading. Stoch Process Appl 11:215–260
Heston SL (1993) A closed-form solution for options with stochastic volatility with applications to bond and currency options. Rev Financ Stud 6(2):327–343
Heston SL, Nandi S (2000) A closed-form GARCH option valuation. Rev Financ Stud 13:585–625
Hsieh K, Ritchken P (2005) An empirical comparison of GARCH option pricing models. Rev Deriv Res 8:129–150
Huang SF (2014) A modified empirical martingale simulation for financial derivative pricing. Commun Stat Theory Methods 43(2):328–334

Huang SF, Tu YT (2014) Asymptotic distribution of the EPMS estimator for financial derivatives pricing. Comput Stat Data Anal 73:129–145
Jackwerth J (2000) Recovering risk aversion from option prices and realized returns. Rev Financ Stud 13:433–451
Jacod J, Shiryaev A (1998) Local martingales and the fundamental asset pricing theorems in the discrete-time case. Financ Stochast 2:259–273
Kreps DM (1981) Arbitrage and equilibrium in economics with infinitely many commodities. J Math Econ 8:15–35
Lindner AM (2009) Continuous time approximations to GARCH and stochastic volatility models. In: Andersen TG, Davis RA, Kreiss JP, Mikosch T (eds) Handbook of financial time series. Springer, Heidelberg, pp 481–496
Lucas RE (1978) Asset prices in an exchange economy. Econometrica 46:1429–1445
Mainardi F, Rogosin S (2006) The origin of infinitely divisible distributions: from de Finetti's problem to Lévy-Khintchine formula. Math Methods Econ Financ 1:37–55
Menn C, Rachev ST (2005) A GARCH option pricing model with α-stable innovations. Eur J Oper Res 163:201–209
Mercuri L (2008) Option pricing in a garch model with tempered stable innovations. Financ Res Lett 5(3):172–182
Monfort A, Pegoraro F (2012) Asset pricing with second-order Esscher transforms. J Bank Financ 36(6):1678–1687
Ortega JP (2012) GARCH options via local risk minimization. Quant Financ 12(7):1095–1110
Pratt JW (1964) Risk aversion in the small and in the large. Econometrica 32:122–136
Rosenberg JV, Engle RF (2002) Empirical pricing kernels. J Financ Econ 64:341–372
Rubinstein M (1976) The valuation of uncertain income streams and the pricing of options. Bell J Econ 7:407–425
Rubinstein M (1985) Nonparametric tests of alternative option pricing models using all reported trades and quotes on the 30 most active CBOE option classes from August 23, 1976 through August 31, 1978. J Financ 40:455–480
Schachermayer W (1992) A Hilbert space proof of the fundamental theorem of asset pricing in finite discrete time. Insur Math Econ 11(4):249–257
Schoutens W (2003) Lévy processes in finance: pricing financial derivatives. Wiley, New York
Schroder M (2004) Risk neutral parameter shifts and derivatives pricing in discrete time. J Financ 59:2375–2402
Siu TK, Tong H, Yang H (2004) On pricing derivatives under GARCH models: a dynamic Gerber-Shiu approach. N Am Actuar J 8:17–31
Walker JS (1996) Fast fourier transforms, 2nd edn. CRC Press, Boca Raton
Yuan ZS, Chen GM (2009a) Strong consistency of the empirical martingale simulation option price estimator. Acta Math Appl Sin (English Series) 25(3):355–368
Yuan ZS, Chen GM (2009b) Asymptotic normality for EMS option price estimator with continuous or discontinuous payoff functions. Manag Sci 55:1438–1450

Empirical Performances of Discrete Time Series Models 4

Having in mind the previously developed models (Chap. 2) and pricing approaches (Chap. 3), this chapter intends to provide the reader with an empirical comparison of them. Even though part of the arguments in this book can seem rather theoretical, the underlying intention behind all of this is to provide a pricing technology that is based on realistic models of financial markets. In this chapter, we will be using a dataset of European options on the S&P500 to test and compare those models. We will not only compare the time series approaches that have been discussed earlier, but compare them to various "standards" of the industry, namely the calibrated and estimated versions of the Heston (1993) model (or at the very least their Heston and Nandi (2000) discrete time counterparts). By doing so, we aim at putting a greater emphasize on a key point most of quantitative analysts around the world are familiar with: rough scales of magnitude for option pricing errors that anyone could expect from a good option pricing model. A large part of the literature diagnoses "fair performances" to various option pricing models. What does it mean and how do those models compare to a properly calibrated standard model?

We will thus first discuss the dataset we rely on, and provide the reader with essential intuitions necessary to the understanding of what is empirically at work in option datasets. Next we focus on the Heston and Nandi (2000) model: using either calibration or time series estimation, we provide measures gauging the pricing error gap between those two ways of putting figures on model parameters. This will help us to understand how this time series approach to option pricing can be useful to the industry, measuring how close the option pricing errors can be to a calibration. This point will then be developed, jointly with a thorough analysis of the parameters estimation results when dealing with GARCH-related time series models with option pricing as an aim.

C. Chorro et al., *A Time Series Approach to Option Pricing*,
DOI 10.1007/978-3-662-45037-6_4

4.1 Historical Dynamics of Option Prices

The empirical estimates presented in this book are based on a dataset of European options based on the S&P500. Such vanilla options have a strong liquidity: they are thus less susceptible to be affected by pricing noise, such as non-synchronous trading prices or strike specific behaviors. Such prices are hence an interesting benchmark for option pricing models: to gauge the quality of an option pricing model, one needs market prices that are as liquid as possible in order that the model parameters do not try to explain what they cannot—frictions, fat fingers or lagging prices. Now, the potential use of the time series approach presented in this book is far from just computing the price of such options. First, market prices already exist and we do not need a model there. Then, this time series approach is clearly well designed to compute a kind of "fundamental" price of illiquid options. It can also provide estimates for prices of options when no organized market pre-exists, as the only necessary ingredient is a dataset of returns for the underlying asset. Still, to be able to use this approach, we need to check at least whether it provides an accurate description of an option market for which market prices exist. This chapter will provide the reader with such a comparison exercise based on a dataset of S&P500 options. When the modeling approach would deliver model-implied prices that match the observed ones, this would increase the likelihood that it can accurately estimate the price of illiquid options.

The rest of this section will now describe the essential features of our option dataset. This dataset starts on the 7th of January 2009 and ends on the 29th of June 2011. It thus excludes the liquidity squeeze at the end of 2008, following the burst of financial markets on Lehman Brothers' default as well as the European debt crisis that happened at the beginning of August 2011. Such periods would have been inappropriate given our liquidity needs. Even though, this dataset still includes various market episodes, such as the February 2009 equity market bottom and the summer 2010 range trading period. Those episodes are important features when selecting an appropriate dataset, as it should include periods covering various market conditions, as long as liquidity remains in line with our needs. These two types of episodes will require the model to deal equally with various scales of risk premia, and thus with stochastic discount factors of various shapes. In order to maintain a liquidity level as strong as possible, we only retain closing prices on Wednesdays. This is a very usual way to deal with such a dataset in the existing literature.[1]

The Table 4.1 presents the number of option prices over the complete dataset, with a breakdown that depends on two important dimensions of any option dataset:

– The time to maturity, that is defined as the difference between the maturity date T of the option and the current date t. This time to maturity thus measures (on

[1] See e.g. Heston and Nandi (2000), Christoffersen et al. (2006) and Barone-Adesi et al. (2008).

Table 4.1 Number of option contracts per moneyness and time to maturity (in months) used in the dataset (January 2009–June 2011)

Number of available option contracts										
Maturity/Moneyness	0.8	0.84	0 .89	0.93	0.98	1.02	1.07	1.11	1.16	1.2
Maturity ≤ 3	45	97	99	97	106	98	96	98	87	80
$3 \leq$ Maturity ≤ 6	43	88	88	90	96	90	81	90	78	73
Maturity ≥ 6	78	165	171	188	189	201	162	177	138	164

a yearly basis) how many years will elapse until the option settlement. A time to maturity of 0.5 is thus a time to maturity of 6 months.

- The moneyness of each option is equal to the ratio between the strike price and the current price of the underlying asset. This ratio measures somewhat the likelihood to see the option exercised at its maturity: in the case of European call options that are our focus here, with a value for this ratio greater than 1 the current probability that the option will be exercised at its maturity is small, unlike the opposite case for which the ratio is below 1.

The total number of option prices in our dataset is 3,353. It covers various maturities and moneyness values. The time to maturity values vary from 7 days to 1 year. The option contracts selected for each date in our sample are the quarterly contracts, that is every March, June, September and December contracts within 1 year. Those contracts are the most liquid ones and thus the most fitted instruments for our purpose. Monthly contracts could also be used, but the history over which those contracts trade is usually much smaller than for the big quarterly ones, not to mention their liquidity. Coming to the moneyness values, we focus on a range of 0.8 to 1.2, as again, that is where we will obtain the best liquidity. Still, such moneyness values are offering an interesting perspective on potential extreme movements in the equity market. Buying a put option with a moneyness of 0.8 consists in buying an insurance policy against a 20 % decrease in the underlying asset—within less than 1 year. Such an event is consistent with the 2008 market crash.

When dealing with option datasets, drawing descriptive statistics on option prices does not make much sense. The first reason is that a given option contract at time t_0 will have a different time to maturity and moneyness at time $t_0 + t$: although the option contract is the same—the strike price and the maturity are unchanged through the life of the option—the nature of the contract can be very different at those two dates because of the effect of time and because the underlying price is evolving. When the underlying asset is worth 1,000$ and the strike price is 800$, the call option price will be mechanically around 200$, the difference between these two.[2] On the contrary, when the underlying asset price is around 700$, the value of

[2]The value of an option is made of two different pieces. First, the intrinsic value that is the value of an immediate exercise of the option, either 0 or the difference between the current price of the underlying asset and the strike price in the case of a call option. The second piece is the time value,

this option is close to zero, for similar reasons. The movements of the underlying asset are thus perturbing any analysis of the fundamental value of the option: in order to get rid of this effect, the analysis of options is usually based on the value of uncertainty—that is the level of volatility implied by the option price.

When it comes to option prices, there are two main kinds of volatilities at work. Using the notations of Sect. 3.1, suppose that the asset log-returns data generating process under an equivalent martingale measure $\mathbb{Q}$ is of the following kind:

$$Y_t = r + m_t^{\mathbb{Q}} + \sqrt{h_t^{\mathbb{Q}}} z_t^{\mathbb{Q}} \tag{4.1}$$

where the $(z_t^{\mathbb{Q}})_{t\in\{1,\dots,T\}}$ are i.i.d random variables with mean 0 and variance 1 under $\mathbb{Q}$ and where $(m_t^{\mathbb{Q}})_{t\in\{1,\dots,T\}}$ is the predictable risk neutral excess of return process. In such a case, $\sqrt{h_t^{\mathbb{Q}}}$ is the risk neutral volatility, that is the conditional standard deviation of returns under the risk neutral distribution $\mathbb{Q}$. In a discrete time version of the Black and Scholes (1973) economy, $m_t^{\mathbb{Q}} = -\frac{h_t^{\mathbb{Q}}}{2}$, the $z_t^{\mathbb{Q}}$ are i.i.d standard Gaussian random variables and $h_t^{\mathbb{Q}} = \sigma^2$ is constant[3]:

$$Y_t = r - \frac{\sigma^2}{2} + \sigma z_t^{\mathbb{Q}}. \tag{4.2}$$

With these settings, option prices are known to be functions of the risk neutral volatility process $\sigma_t = \sqrt{h_t^{\mathbb{Q}}}$. In particular, in the Black and Scholes (1973) case, option prices are computed through a closed-form expression: for a European call option with a strike price K and maturity T, the price at time t is given by the so-called Black and Scholes formula:

$$C(t, T, K, S_t, r, \sigma) = S_t N(d_1) - Ke^{-r(T-t)} N(d_2) \tag{4.3}$$

$$d_1 = \frac{log(\frac{S_t}{K}) + (r + \frac{\sigma^2}{2})(T-t)}{\sigma\sqrt{T-t}} \tag{4.4}$$

$$d_2 = d_1 - \sigma\sqrt{T-t} \tag{4.5}$$

where S_t is the value at time t of the underlying asset, r is the risk-free rate and $N(x)$ is the distribution function of a standard Gaussian distribution, that is

$$N(x) = \int_{-\infty}^{x} \frac{1}{\sqrt{2\pi}} e^{-\frac{y^2}{2}} dy. \tag{4.6}$$

that is the value coming from the fact that the option is not exercised today. Usually, this time value is low.

[3]In this case, the historical and the risk neutral volatilities coincide (see Proposition 3.2.2).

This expression can be used in two ways: the first way—and probably the most natural—is to compute option prices using the option contract specifications and an estimation of the volatility σ. Volatility is indeed the only unobserved parameter in the Black and Scholes formula, and it thus requires an estimation strategy. When this volatility estimate is obtained from the time series evolution of returns, such an approach is referred to as the "direct modelling approach" in Bertholon et al. (2008). Useful as this exercise may have been in the past, many option markets are now liquid enough for a second type of use of this formula, that is extracting the volatility from existing option prices. This "implied volatility" is obtained by inverting the Black and Scholes formula given the option prices observed in the market.[4] Thus, to compute such an implied volatility we should first rely on a function that computes Black and Scholes option prices. Such a function can be:

```
bs<-function(s_h,T,K,S,r,type="CALL"){

# s_h is the square root of h, that is the volatility
# T is the time to maturity (not the maturity here)
# K is the strike price
# S is the current value of the underlying
# r is the risk free rate


d1=log(S/(K*exp(-r*T)))/(s_h*sqrt(T))+1/2*s_h*sqrt(T)
d2=d1-s_h*sqrt(T)

P=S*pnorm(d1)-K*exp(-r*T)*pnorm(d2)

# When the option contract is a PUT, just change type from "CALL"
# into "PUT" and obtain the price from the call-put parity

if (type!="CALL"){
  P=P-S+K*exp(-r*T)
}
return(P)
}
```

[4]Computing the derivative of Eq. (4.3) with respect to σ, we obtain the so-called Véga (sensitivity of the price with respect to the volatility) that is strictly positive: the call option price is a strictly increasing (and thus invertible) function of the volatility. Given the integral involved in the formula, there is no closed-form expression to such an inverse function. It has to be numerically approximated. One possibility (that is used in the following) is to minimize the quadratic difference between the theoretical price and the actual market price. We also mention that very accurate approximations for N^{-1} exist (see Cody 1969) and are implemented in R via the *qnorm* function of the *stats* package.

Based on this function, it is now possible to create a function that computes the quadratic difference between the price given by the model and the actual market price:

```
invert_bs<-function(s_h,P,T,K,S,r,type="CALL"){

# P is the actual market price

error=P-bs(s_h,T,K,S,r,type)
return(error^2)
}
```

Now, a minimizer can be used to find the volatility σ that makes the Black and Scholes option price as close as possible to the market price of the option. Here is a dummy example of such a computation:

```
# An imaginary call option contract:
S=100 # Underlying asset price
K=80 # Strike price
T=0.3 # Time to maturity
r=0.02 # Risk free rate
P=23.4 # The market price of this option

# When optimizing for a single parameter, R favors "optimize" instead of
#"optim"

optimize(f = invert_bs, interval =c(0,1) , lower = 0,
         upper = 1, maximum = FALSE,
         tol = .Machine\$double.eps^0.25,P,T,K,S,r)
```

Should this Black and Scholes modeling approach accurately describe the evolution of financial markets and when inverting $C(t, T, K, S, \sigma)$ using quoted option prices, σ should not be a function of the strike price used. Graphically, the implied volatilities for a given time to maturity and various strikes should be a flat horizontal line. For such a theoretical case, the implied volatility coincides with the volatility of the underlying asset returns in Eq. (4.2).

Using the previously mentioned dataset, we now graph in Fig. 4.1 these implied volatilities as a function of the moneyness and of the time to maturity. The figure presents the average implied volatility surface in our sample. Obviously, this surface is neither flat as a function of the moneyness nor of the time to maturity: when looking at a given time to maturity, the shape is convex and somewhat asymmetric. It is usually referred to as "the implied volatility smile", given its shape. What is more, for a given moneyness the implied volatility is an increasing function of the time to maturity. For those two reasons, implied volatilities can not be mixed with the risk neutral volatility: the risk neutral volatility is the volatility of the underlying asset returns under the risk neutral probability. It thus does not depend on the time to maturity or the moneyness of the option. Implied volatilities are thus just another way to look at the option prices: given the one to one mapping between prices and volatilities through the Black and Scholes formula, looking at prices or at implied volatilities is exactly the same. The major difference between prices and

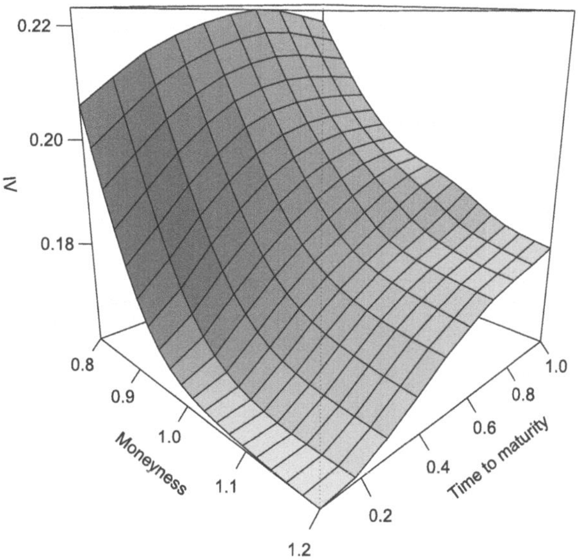

Fig. 4.1 Average option implied volatility surface obtained from the S&P500 options over 2009–2012

volatilities is that implied volatilities are presented in a way that gets rid of the dependence on the underlying asset value. For example, options on the German DAX and on the US S&P500 have very different prices primarily because of the very different levels of their underlying assets: when the S&P 500 evolved in 2011 around 1,300$, the DAX ranged around 7,000 Euros. Given that the price of a call option is somewhat a weighted difference between the underlying asset price and the strike price, the difference in option prices can be significant. Now, when looking at implied volatilities, we naturally get rid of this scale problem. Volatilities are comparable quantities, and thus make it possible to analyze option prices for several underlying assets.

Another reason explaining why analyzing option markets through the prism of implied volatilities was maintained in the industry is that it makes it possible to understand part of the risk neutral mechanics: thinking in terms of Black and Scholes implied volatilities boils down to thinking of option markets using the Gaussian distribution hypothesis as a benchmark. The very fact that implied volatilities depart from the Black and Scholes expectation of a straight line implies that the world is not Gaussian. Gauging how non-Gaussian this world is can be done through an analysis of the implied volatility surface. A very first way to look at that is to observe that implied volatilities are time varying, as presented in Fig. 4.2: implied volatilities for a fixed moneyness and time to maturity exhibit significant variations through time. This behavior is very similar to what is observed under the historical measure, and suggests that the risk neutral conditional distribution itself is time varying. Beyond that, another popular way of understanding implied volatilities dynamics is to analyze the factors at work behind implied volatilities datasets.

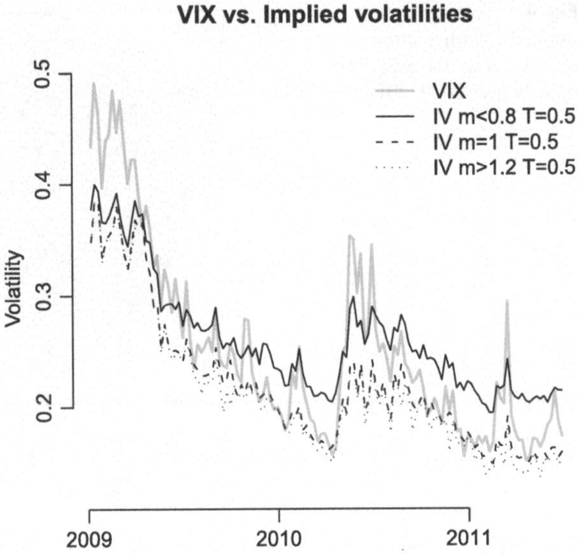

Fig. 4.2 Evolution of various implied volatilities with a 6 months time to maturity compared to the VIX index over January 2009–June 2011

Given the important correlation that we can expect from implied volatility surfaces by construction, there is an important information overlap between each series.[5] When there is such an information overlap, it is usually possible to account for the behavior of the dataset using a number of unobservable quantities whose number is much smaller than the number of series in the dataset. This is called a factor representation: a few series explain most of the evolution of the investigated dataset, and thus will help us to understand the dynamics at work.

The joint behavior of implied volatilities is analyzed under the factor model specification. A factor model's formulaic appearance is identical to that of a simple regression model, but a different interpretation is attached to each element. The elements' familiar names under the simple regression setting is abused in this first introduction to factor models to obtain an intuitive grasp of the concept[6]:

- The *dependent variable* is the daily changes in the logarithm of implied volatilities.
- The *independent variables* are then the common components, also known as *factors*, which are assumed to be unobservable.

[5]By "series" here, we refer to a dataset of implied volatilities for a given moneyness and a fixed time to maturity, across the dates of our sample.

[6]We refer the reader to Chamberlain and Rothschild (1983) and Bai and Ng (2002) for an introduction to factor models that have become increasingly important in both finance and macroeconomics.

- The *error terms* capture the idiosyncrasies of the implied volatilities, i.e. strike and maturity specific variables that explain variance in implied volatilities changes.
- The coefficients associated with each factor are known as the *loadings*.

Thus, a factor model consists of observed data, factors (unobservable), their loadings, and idiosyncratic terms. An *approximate factor model* with r factors has the following representation: $\forall i \in \{1, \ldots, N\}$, $\forall t \in \{1, \ldots, T\}$[7]

$$X_{m_i,\tau_i,t} = \lambda'_{m_i,\tau_i} F_t + e_{m_i,\tau_i,t} \tag{4.7}$$

where

- $X_{m_i,\tau_i,t}$ is the daily change between $t-1$ and t in the logarithm of the implied volatility for the ith set of moneyness and time to maturity (m_i, τ_i) at time t.
- F_t is the $r \times 1$ vector of unobservable common factors.
- λ_{m_i,τ_i} is the $r \times 1$ vector of factor loadings.
- $e_{m_i,\tau_i,t}$ is the idiosyncratic component.
- $'$ denotes the complex conjugate transpose of the matrix.

If we denote $X_t = (X_{m_1,\tau_1,t}, \ldots, X_{m_N,\tau_N,t})'$, $\Lambda = (\lambda_{m_1,\tau_1}, \ldots, \lambda_{m_N,\tau_N})'$, $F = (F_1, \ldots, F_T)'$ and $e_t = (e_{m_1,\tau_1,t}, \ldots, e_{m_N,\tau_N,t})'$, the model (4.7) is represented in the matrix form

$$X = \Lambda F' + e \tag{4.8}$$

where $X = (X_1, \ldots, X_T)$ and $e = (e_1, \ldots, e_T)$. The following assumptions are imposed:

1. The factor F_t and the errors e_t are uncorrelated.
2. The covariance matrix of e_t is not necessarily diagonal, allowing for serial correlation and heteroskedasticity, but the degree of correlation between the idiosyncratic components is limited, i.e. the largest eigenvalue of the covariance matrix of e_t is assumed to be bounded.

The estimation of the common factors is now accomplished via a dimension-reduction tool known as Principal Component Analysis (PCA), which aims to select the salient factors that explain variances in the data. Estimates of Λ and F are

[7]N being the number of different (moneyness,time to maturity) pairs in our dataset and T the number of dates.

obtained by solving a minimization problem (depending on the initial fixed number of factors r), set up to yield the least error-sum-squared:

$$V(r) = \min_{\Lambda, F} \frac{1}{NT} \sum_{i=1}^{N} \sum_{t=1}^{T} (X_{m_i,\tau_i,t} - \lambda'_{m_i,\tau_i} F_t)^2$$

subject to the normalization of either $\frac{F'F}{T} = I_r$ and $\Lambda'\Lambda$ is diagonal, or $\frac{\Lambda'\Lambda}{N} = I_r$ and $F'F$ is diagonal. After some tedious linear algebraic manipulations, we can demonstrate that the estimated factor matrix $\hat{F}$ is essentially the eigenvectors corresponding to the r largest eigenvalues of the matrix $X'X$.

Using our dataset of options, we first extract implied volatilities before applying the preceding technique. Then, for each day, we compute the implied volatilities for specific moneyness and time to maturity values, using a kernel regression as advised in Cont and da Fonseca (2002).[8] One key issue with implied volatility surfaces is that to be able to study their dynamics, we need fixed moneyness and time to maturity values across time. The problem is that option contracts have by essence a constantly decreasing time to maturity—as time passes by—and a changing moneyness as the underlying asset evolves. If there is a factor structure in implied volatility datasets, it should be observable once variations in implied volatilities are comparable with each other: this is why we need to interpolate implied volatilities to maintain constant time to maturity and moneyness at any time. Thus, observing a given option contract will be of little help when it comes to trying to understand the evolutions of volatility surfaces. To do so, we need to interpolate as precisely as possible the implied volatilities for a set of given moneyness and time to maturity values. Several interpolation methods can be used here:

- The most classical method would simply interpolate the target point using the closer two points of the surface. Naturally, when one of the two options is missing, such a method is simply impossible to use.
- Cubic polynomials could also be used, as they proved to be able to mimic the shape of the implied volatility for a given time to maturity (see Shimko 1993). Such methods have still difficulties handling the time to maturity effect and are very sensitive to outliers in the dataset.
- Finally, non-parametric estimation methods are especially well-designed to estimate implied volatility surfaces. Let $\sigma^{IV}(m, \tau)$ be the implied volatility at a given date, for a time to maturity equal to τ and a moneyness m. Suppose that we have a set of implied volatilities for this day (corresponding to a set of moneyness/time to maturity values $(m_i, \tau_j)_{(i,j)\in I\times J}$) and we wish to deduce $\sigma^{IV}(m, \tau)$ from it. Cont and da Fonseca (2002), along with Ait-Sahalia and

[8] See Eq. (7), page 49 of Cont and da Fonseca (2002).

Lo (1998), propose to use a Nadaraya-Watson estimator. This non-parametric estimator writes:

$$\sigma^{IV}(m,\tau) = \frac{\sum_{j\in J}\sum_{i\in I} K(m-m_i,h_m)K(\tau-\tau_j,h_\tau)\sigma^{IV}(m_i,\tau_j)}{\sum_{j\in J}\sum_{i\in I} K(m-m_i,h_m)K(\tau-\tau_j,h_\tau)} \tag{4.9}$$

where $K(x,h)$ is the density function of a centered Gaussian distribution with a variance equal to h.[9] This type of estimator combines the different implied volatilities $\sigma^{IV}(m_i,\tau_j)$ in our sample depending on how m_i and τ_j are close to m and τ, the targeted moneyness and time to maturity. It is thus a simple way to decide on proper weights to be put on observations when thinking in terms of interpolation. The following piece of code performs the computation of such a function:

```
kernel_reg<-function(ki,Ti,x,k,T)
{

# Kernel regression to estimate the implied volatility value for
# the moneyness ki and the time to maturity Ti, given the dataset
# of implied volatilities x and their corresponding moneyness k
# and time to maturity T values

# Rule of thumb for  smoothing parameters
h1=1.8*sd(k)/(length(k)^(2/5))
h2=1.8*sd(T)/(length(T)^(2/5))

temp=dnorm(x,k-ki,h1)*dnorm(x,T-Ti,h2)
ratio1=temp*x
ratio2=temp
ratio=sum(ratio1)/sum(ratio2)
return(ratio)
}
```

We now perform a PCA over the dataset of daily changes in the logarithm of implied volatilities with constant time to maturity and moneyness obtained from using the previous non-parametric methodology: we first compute the implied volatilities for a given day using the available option prices and then interpolate those volatilities for fixed values of the time to maturity and of the moneyness. Using the daily log-variations of all these interpolated series, we then compute the matrix $X'X$ (see Eq. (4.8)) from which we obtain the eigenvalues and eigenvectors

[9] h is actually working as a smoothing parameter: the bigger this parameter is and the smoother is the function. An usual way to set the smoothing parameter h associated to a dataset x (here the moneyness or the maturity values) is to set it to be equal to $\frac{1.8\sigma_x}{n^{2/5}}$, where n is the total number of observations in x and σ_x the corresponding empirical standard deviation.

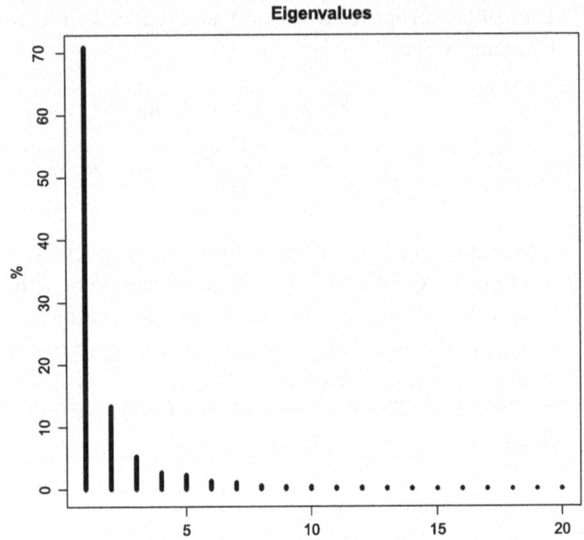

Fig. 4.3 First 20 eigenvalues divided by the sum of all the eigenvalues obtained from the matrix $X'X$ (see Eq. (4.8)) computed from the daily changes in implied volatilities

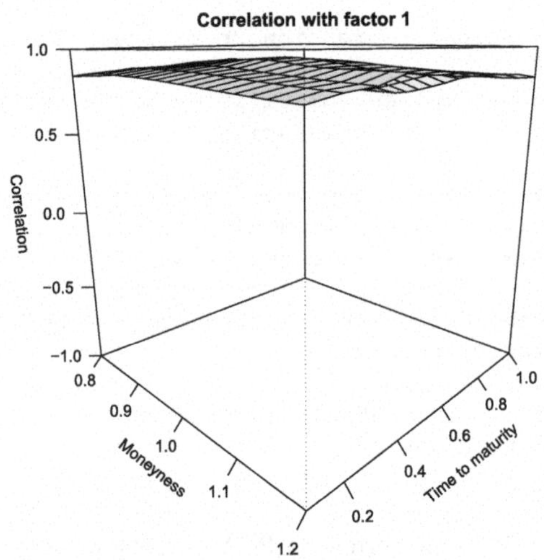

Fig. 4.4 Correlation between the implied volatility surface and the first factor obtained from a PCA analysis

necessary to the PCA. The eigenvalues are presented in Fig. 4.3. A subset of three factors accounts for more than 90 % of the variations of our dataset: with only three factors, we get to know much of the time variations of volatility surfaces. We identify these factors by computing the correlations between our dataset and the estimated factors. We chart these correlations as a function of the time to maturity and of the moneyness for the first (Fig. 4.4), the second (Fig. 4.5) and the third factor (Fig. 4.6).

Similarly to Cont and da Fonseca (2002) and despite the fact that their sample and methodology differ from ours, we still diagnose that three factors accurately describe the evolution of the implied volatility surface. Factor 1 is positively related

a timely manner—one of the must have when selecting methods for their potential application in the industry. This is why we now turn our attention toward the Heston and Nandi (2000) model that combines the two good features of GARCH models and of the Heston model: its parameters can be classically estimated by conditional maximum likelihood and its characteristic function is still an exponential affine function of the state variables, making the pricing of European options fast.

As seen in Sect. 2.7, this model can be written as follows:

$$\begin{cases} Y_t = log\left(\frac{S_t}{S_{t-1}}\right) = r + \lambda_0 h_t + \underbrace{\sqrt{h_t} z_t}_{\varepsilon_t}, \\ h_t = a_0 + a_1 (z_{t-1} - \gamma\sqrt{h_{t-1}})^2 + b_1 h_{t-1}, \\ S_0 = s \in \mathbb{R}_+. \end{cases} \tag{4.12}$$

where the z_t are, under the historical probability $\mathbb{P}$, i.i.d $\mathcal{N}(0,1)$ that generate the information filtration. From the relation

$$E_{\mathbb{P}}[h_{t+1}|\mathcal{F}_t] = a_0 + a_1 + (a_1\gamma^2 + b_1)h_t,$$

it can be proved, using the same reasoning as in Sect. 2.2.3, that a unique second order stationary solution to (4.12) exists if and only if $a_1\gamma^2 + b_1 < 1$. The average level of volatility is now a combination of all the volatility parameters, as

$$\mathbb{E}[h_t] = \frac{a_0 + a_1}{1 - a_1\gamma^2 - b_1} \tag{4.13}$$

and the economic interpretation of each parameter is therefore slightly more difficult than in the continuous time model of Heston (1993): the long term level of the variance (4.13) is an increasing function of any volatility parameter (the others being constant). The persistence of the variance process is jointly driven by $a_1\gamma^2$ and by b_1. Finally, the returns-to-volatility feedback part of the process is mainly driven by the γ parameter: as a_1 is naturally always expected to be positive in order to preserve the positivity of the variance process, when γ is positive, the Heston and Nandi model is able to generate an equity-like leverage effect. Hence, this discrete time model is able to generate the same type of volatility behaviors as the Heston (1993) model: the variance process is time-varying, persistent and able to generate leverage effects. This specification does however something more than the continuous time model: as there is now a single source of randomness, that is z_t, the parameters can be estimated using the conditional maximum likelihood presented in Sect. 2.6.1. We denote by $\theta = (a_0, a_1, b_1, \gamma, \lambda_0)$ the parameters of the model.[14]

[14]The purpose of this chapter is to provide the reader with a guide on how to use an option pricing model in the standard way: by standard, we mean that before computing option prices, we need to find a set of values for the parameters that fits the distribution of the returns. This step should be an essential one and is often overlooked, as rather liquid option markets made this step useless.

Remarking that the conditional distribution of Y_t given $\mathcal{F}_{t-1}$ is a $\mathcal{N}(r + \lambda_0 h_t, h_t)$, the conditional log-likelihood based on the observations $(y_1, \ldots, y_T)$ can then be written (see Remark 2.6.1) as follows:

$$L_T(y_1, \ldots, y_T \mid \theta) = \sum_{t=1}^{T} -\frac{1}{2}\left(log(2\pi) + log(h_t) + \frac{(y_t - r - \lambda_0 h_t)^2}{h_t}\right). \tag{4.14}$$

A reliable set of estimated parameters $\hat{\theta}_T$ is therefore obtained by numerically maximizing this latter expression with respect to the parameters. It has to be done numerically as no closed-form expressions for the estimators of the parameters can be obtained, due to the recursive form of the log-likelihood function. The following R function shows how to compute the log-likelihood for a given set of parameters:

```
loglik_heston<-function(para,x,r){

# "para" is a vector containing the parameters over
# which the optimization will be performed
# "x" is a vector containing the historical returns on the underlying asset
# r is the risk-free rate, expressed here on a yearly basis

 a0=para[1] # the variance's intercept
 b1=para[2] # the persistence parameter
 a1=para[3] # the autoregressive parameter
 gamma=para[4] # the leverage parameter
 lambda0=para[5] # the risk premium

# the log-likelihood is initialized at 0
loglik=0

# The first value for the variance is set to be equal to its long term value
h=(a0+a1)/(1-b1-a1*gamma^2)

# The next for loop recursively computes the conditional variance,
# risk premium and the associated density

for (i in 1:length(x)){
# The conditional log-likelihood at time i is:
    temp=dnorm(x[i],mean=r[i]/250+lambda0*h,sd=sqrt(h),log=TRUE)
    # The full log-likelihood is then obtained by summing up
    # those individual log-likelihood
    loglik=loglik+temp
    # The epsilon is then computed:
    eps=x[i]-(r[i]/250+lambda0*h)
```

When vanilla option markets are liquid, the parameters under the pricing measure can be recovered from option prices—a method usually referred to as "calibration". However, liquidity traps and the need to price options on new underlying assets make this calibration step inadequate from time to time. What is more, we should mention that the understanding of the economic intuitions hidden behind an option pricing model and the study of historical patterns should make the quantitative analysts around the world more cautious about the ability of pricing models to be reliable under any circumstances.

```
    # An the conditional variance is updated as well
    h=a0+a1*(eps/sqrt(h)-gamma*sqrt(h))^2+b1*h
  }

# R provides minimizers, that  is why the maximum likelihood is
# obtained by trying to minimize -loglik
return(-loglik)
}
```

Using a dataset of returns and of risk-free rates, the previous function needs to be minimized with respect to its parameters to obtain maximum likelihood estimates for the parameters. This can be done using different optimizers in R. Here, we use the R multipurpose optimizer `optim`.[15] Given a dataset of returns and of risk-free rates, the estimation is thus performed by running the following code line[16]:

```
optim(initial_parameters,loglik_heston,,returns,risk_free_rates)
```

where `initial_parameters` is the vector of the initial parameters from which the optimization will start, `loglik_heston` is the log-likelihood function previously detailed, `returns` is a vector of returns while `risk_free_rates` contains the corresponding risk-free rates. The initial parameters selection is obviously a very important step: when badly chosen values are selected, the outcome of the optimization is very likely to be wrong, given the complicated shape such functions can take. Given the dimension of the optimization problem, it is impossible to chart its shape in a 3-dimension graph: the modeling experience and the user's intuitive understanding of the parameters are key here. Browsing through the empirical literature can also be very useful.

Table 4.2 and Fig. 4.7 present descriptive statistics around the rolling estimation results of the parameters of the model. Those results are obtained by performing estimation over periods of 4,000 trading days, that is a bit less than 16 years. The dataset length must be large enough to contain rare extreme events as well as quieter periods of trading. In the meantime, given that the model uses a limited number of fixed parameters, it will not be able to accommodate large shifts in the evolution of financial returns. By performing this rolling estimation, we will then be able to compute option prices in a realistic manner: using the last 4,000 returns until time $t_0 - 1$, once the parameters have been estimated, we will be able to characterize the risk neutral distribution and finally to compute option prices at time t_0. We perform the rolling estimation using daily data. We estimate the parameters for every blocks of 4,000 returns ending on a Wednesday between the 7th of January 2009 and the 26th of June 2011, the same dataset as in the previous section.

[15]An interested reader should browse the documentation associated with the following optimizers: `optim`, `nlm` and `nlminb`. Each of them provides optimization solutions depending on the nature of the optimization problem the user is faced with.

[16]The order of the arguments of the `optim` function has been changing over the years depending on the version of `R`. We advise the interested reader to look at the documentation associated with the function by typing `?optim` in the software console.

Table 4.2 Estimated parameters for the Heston and Nandi model

	a_0	b_1	a_1	γ	λ_0
Average	4.29E−07	0.662	1.51E−06	462.6	0.64
Standard deviation	5.67E−08	0.002	7.48E−09	0.3	0.33
Quantile 2.5 %	3.49E−07	0.656	1.50E−06	462.1	0.20
Quantile 97.5 %	5.20E−07	0.665	1.52E−06	463.0	1.11

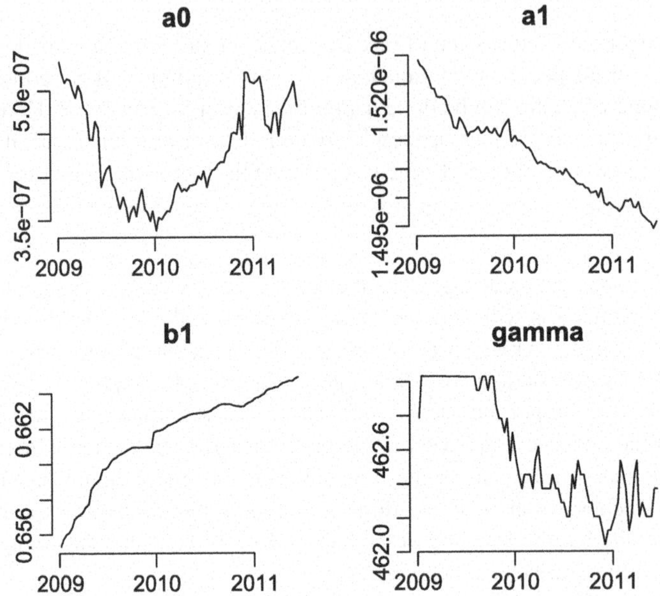

Fig. 4.7 Rolling estimation of the Heston and Nandi model over 2009–2011

Table 4.2 presents the parameter estimation results obtained. Again, the economic interpretation is not straightforward. However, a couple comments can still be made. First of all, the estimated parameters are found to be very close to those obtained in the original version of the Heston and Nandi (2000) article, using a different dataset. This stability feature is also observable in the quantiles computed out of our rolling estimation scheme: for most of the estimated parameters, the values obtained exhibit a very limited time variation. For example, b_1 ranges between 0.656 and 0.665 95 % of the time. Similar cases can be made for a_1 and γ. Given that a_0 somewhat drives the long term average volatility (a_1, b_1 and γ being quiet stable), and that this long term average is subject to changes in regimes, this parameter shows a higher degree of variation. A similar situation is observed in the case of λ_0: the risk premium varies a lot over the sample that we use. Given that this sample incorporates various market episodes, the fact that some of them are incorporated in the rolling sample or not can change the estimated risk premium. These results should provide an interested reader with two intuitions: first, estimated parameters can be affected by the selected sample, as shown by Table 4.2 and

Fig. 4.7. Second, this time variation in the case of the Heston and Nandi (2000) model is *very* limited. The observed variations are mostly related to changes in the financial conditions: Fig. 4.7 shows rather clearly how a_0 decreases over the 2009 sharp recovery, to increase again afterward. The drift that is exhibited in the b_1 and a_1 cases is also a sign that the model is changed by the estimation methodology to fit changing conditions, suggesting that volatility shocks are increasingly persistent over the sample. Should the model be overfitting, the time series at hand and the evolution of those parameters would look random instead of affected by trends.

Once we know more about the historical dynamics, we can turn to the risk neutral one. For the Heston and Nandi model, the risk neutral dynamics under the Esscher EMM $\mathbb{Q}^{ess}$ is given by the following equations[17] (see Example 3.4.1):

$$Y_t = r - \frac{h_t}{2} + \sqrt{h_t}\xi_t, \quad S_0 = s \tag{4.15}$$

$$h_t = a_0 + a_1(\xi_{t-1} - \underbrace{(\gamma + \lambda_0 + \frac{1}{2})}_{\gamma^*}\sqrt{h_{t-1}})^2 + b_1 h_{t-1} \tag{4.16}$$

where the ξ_t are i.i.d $\mathcal{N}(0, 1)$ under $\mathbb{Q}^{ess}$. This change in the distribution has two important consequences, one affecting returns and one affecting their volatility:

- First of all, as in the Black and Scholes (1973) case, the expected return is changed to equate the risk-free rate.
- Then, the volatility is modified through a change in the γ parameter: under the risk neutral probability, it is now equal to $\gamma^* = \gamma + \lambda_0 + \frac{1}{2}$. This has two main implications. First, given that λ_0 has been found to be positive, we have:

$$\gamma < \gamma^*. \tag{4.17}$$

Hence, under the risk neutral probability, the average volatility in the Heston and Nandi model is higher than in the historical case, given the set of values obtained for the parameters. Next, given that γ^* drives the leverage effect under the risk neutral probability, there is a higher degree of leverage under $\mathbb{Q}^{ess}$. This is somewhat consistent with the skewness premium found in option prices, that is a stronger skewness implied in option prices than in financial returns.[18] Having said that, in the case of a negative risk premium, we obtain exactly the opposite situation: when the risk premium is negative, $\gamma > \gamma^*$, the skewness premium is positive, and the risk neutral volatility is lower than the historical one. This matches the empirical observations: during financial crises, risk premia are in

[17]Remind that in the conditionally Gaussian case, the dynamics implied by the conditional Esscher transform and the extended Girsanov principle (see Sect. 3.3) coincide with the one coming from the LRNVR in Proposition 3.2.2.

[18]On this point, see Bates (1997).

general negative: the returns obtained from stocks (roughly −20 % in 2008) are far below the ones obtained from a risk-free investment (roughly 2–3 % in 2008). During such periods, historical volatility is also known to be higher than the one implied in option prices.

In this respect, the Heston and Nandi (2000) framework is able to replicate several of the known features of the relationship between historical and risk neutral distributions of returns. This is however not enough: the model also needs to be able to replicate the scale of those effects. To judge of that, further numerical analyzes are required: the rest of this section will be dedicated to assessing the performances of this option pricing model when confronted with market data. We perform two types of experiments: first, we provide the reader with the pricing performances obtained when performing a parameter calibration—that is picking the parameters in order to best fit option prices—and then compare them to the pricing performances when first performing the historical estimation of the parameters before using the previously mentioned change in the parameters to derive the risk neutral distribution from the historical one. By doing so, the reader will be able to assess and understand what it takes for a model to provide accurate option prices.

Common to these two last issues, the pricing of options within the Heston and Nandi (2000) model is an essential step. Given that the characteristic function is available in a closed-form expression, we simply use the Carr and Madan (1999) method that is faster and more accurate than the Monte Carlo methods we will use later in the case of models for which such a closed-form expression does not exist. The conditional moment generating function of $log(S_T)$ given $\mathcal{F}_t$ of the Heston and Nandi (2000) model is given by the following expression (see Proposition 3.7.1):

$$\mathbb{G}^{\mathbb{Q}^{ess}}_{log(S_T)|\mathcal{F}_t}(u) = S_t^u e^{A_t + B_t h_{t+1}}, \tag{4.18}$$

where

$$A_t = ru + A_{t+1} + a_0 B_{t+1} - \frac{1}{2} log(1 - 2a_1 B_{t+1}) \tag{4.19}$$

and

$$B_t = \frac{-1}{2} u + b_1 B_{t+1} + \frac{\frac{u^2}{2} - 2a_1 \gamma^* B_{t+1} u + a_1 B_{t+1} (\gamma^*)^2}{1 - 2a_1 B_{t+1}} \tag{4.20}$$

with terminal conditions $A_T = B_T = 0$. Heston and Nandi (2000) provide in their article (see also Sect. 3.7) a semi-closed-form formula for the price of options:

$$\begin{aligned} e^{-r(T-t)} E_{\mathbb{Q}^{ess}}[(S_T - K)_+ \mid \mathcal{F}_t] = \tfrac{S_t}{2} + \tfrac{e^{-r(T-t)}}{\pi} \int_0^{+\infty} \mathcal{R}e \left[\frac{K^{-i\phi} \mathbb{G}^{\mathbb{Q}^{ess}}_{log(S_T)|\mathcal{F}_t}(i\phi + 1)}{i\phi} \right] d\phi \\ -Ke^{-r(T-t)} \left(\tfrac{1}{2} + \tfrac{1}{\pi} \int_0^{+\infty} \mathcal{R}e \left[\frac{K^{-i\phi} \mathbb{G}^{\mathbb{Q}^{ess}}_{log(S_T)|\mathcal{F}_t}(i\phi)}{i\phi} \right] d\phi \right). \end{aligned}$$

This formula however implies the computation of an integral and a discretization scheme for it. In the meantime, this computation will have to be performed for each of the strikes and maturities for which we need to compute option prices. Even though the actual numerical cost of such an integration is not heavy when computing it only a couple of times, this cost will massively increase when we will need to optimize the parameters to fit option prices: in such a case, we will have to perform this computation a very large number of times along the optimization process, resulting into a much slower process.

To deal with such issues, Carr and Madan (1999) developed an approach based on the so-called Fast Fourier Transform (FFT). This method is fast and makes it possible to compute the price of options for a full range of strikes and a given maturity, decreasing massively the required time to price a full option book. We do not provide here a full description of what a FFT is[19]: an interested reader will read Walker (1996) on this point. Carr and Madan (1999) is based on the characteristic function of the final log-price of the risky asset. Let us add the following notations:

$$k = log(K) \tag{4.22}$$

$$s = log(S_T) \tag{4.23}$$

and q_T is the risk neutral density function of $log(S_T)$ under $\mathbb{Q}^{ess}$. Using those notations, the price at time $t = 0$ of a European call option with strike K and maturity T is given by the following formula (see Eq. (3.11)):

$$C_T(k) = \int_k^{\infty} e^{-rT}\left(e^s - e^k\right) q_T(s)ds. \tag{4.24}$$

Such an expression is however not integrable (it tends to the positive constant S_0 in $-\infty$), condition that is necessary to apply the FFT. Carr and Madan (1999) start instead with the following transform of the option pricing formula:

$$c_T(k) = e^{\alpha k} C_T(k) \tag{4.25}$$

---

[19]Roughly speaking, the FFT is an efficient algorithm due to Cooley and Tukey (1965) which allows to compute the sums $F_0, \ldots, F_{N-1}$ where

$$F_k = \sum_{j=0}^{N-1} e^{\frac{-2i\pi jk}{N}} x(j) \tag{4.21}$$

using only $O(Nlog(N))$ operations instead of $O(N^2)$.

where $\alpha > 0$ is chosen in order to make $c_T(k)$ square integrable.[20] The Fourier transform of $c_T(k)$ is then:

$$\psi_T(\omega) = \int_{-\infty}^{\infty} e^{iwk} c_T(k)dk \tag{4.26}$$

that can be expressed in terms of Φ_T, the characteristic function of $log(S_T)$ under $\mathbb{Q}^{ess}$:

$$\begin{aligned}
\psi_T(\omega) &= \int_{-\infty}^{\infty} e^{iwk} \int_{k}^{\infty} e^{\alpha k} e^{-rT}(e^s - e^k) q_T(s)dsdk && (4.27)\\
&= \int_{-\infty}^{\infty} e^{-rT} \int_{k}^{\infty} (e^{(\alpha+iw)k+s} - e^{(\alpha+1+iw)k}) q_T(s)dsdk && (4.28)\\
&= \int_{-\infty}^{\infty} e^{-rT} q_T(s) \int_{-\infty}^{s} e^{(\alpha+iw)k+s} - e^{(\alpha+1+iw)k} dkds && (4.29)\\
&= \int_{-\infty}^{\infty} e^{-rT} q_T(s) e^{(iw+\alpha+1)s} \left(\frac{1}{iw+\alpha} - \frac{1}{1+iw+\alpha} \right) ds && (4.30)\\
&= \frac{e^{-rT}}{\alpha^2+\alpha+i(2\alpha+1)w-w^2} \int_{-\infty}^{\infty} q_T(s) e^{(iw+\alpha+1)s} ds && (4.31)\\
&= \frac{\Phi_T(w - i(\alpha+1))e^{-rT}}{\alpha^2+\alpha+i(2\alpha+1)w-w^2} && (4.32)
\end{aligned}$$

Finally, using the classical Fourier inverse transform, the option price is given by the following expression:

$$\begin{aligned}
C_T(k) &= e^{\alpha k} c_T(k) && (4.33)\\
&= \frac{e^{\alpha k}}{2\pi} \int_{-\infty}^{\infty} e^{-iwk} \psi_T(w)dw && (4.34)\\
&= \frac{e^{\alpha k}}{\pi} \int_{0}^{\infty} e^{-iwk} \psi_T(w)dw. && (4.35)
\end{aligned}$$

Substituting (4.32) in (4.35), we obtain

$$C_T(k) = \frac{e^{\alpha k}}{\pi} \int_{0}^{\infty} e^{-iwk} \frac{\Phi_T(w - i(\alpha+1))e^{-rT}}{\alpha^2+\alpha+i(2\alpha+1)w-w^2} dw. \tag{4.36}$$

[20] In practice we will take $\alpha = 2$, value that correctly works in our empirical study. See Carr and Madan (1999) for a discussion on the choice of this parameter.

In order to use the FFT to obtain an efficient algorithm to compute very quickly $C_T(k)$ for different values of k, we try to express (4.36) as a discrete Fourier transform (4.21). First we truncate the integral

$$C_T(k) \approx \frac{e^{\alpha k}}{\pi} \int_0^a e^{-iwk} \frac{\Phi_T(w - i(\alpha+1))e^{-rT}}{\alpha^2 + \alpha + i(2\alpha+1)w - w^2} dw.$$

Then, using a regular discretization of $[0, a]$ with $N + 1$ equidistant points (the stepsize being $\delta = \frac{a}{N}$), the associated Riemann sum approximation is

$$C_T(k) \approx \frac{e^{\alpha k}}{\pi} \sum_{j=0}^{N-1} e^{-ij\delta k} \psi_T(j\delta)\,\delta. \tag{4.37}$$

Let us consider N values for the log-strike k, regularly spaced in the interval $[-b, b]$:

$$k_l = -b + \lambda l, \forall l \in \{0, \ldots, N-1\}, \tag{4.38}$$

where $\lambda = \frac{2b}{N}$. Thus

$$C_T(k_l) \approx \frac{e^{\alpha k_l}}{\pi} \sum_{j=0}^{N-1} e^{-ij\delta(-b+\lambda l)} \psi_T(j\delta)\,\delta \tag{4.39}$$

$$\approx \frac{e^{\alpha k_l}}{\pi} \underbrace{\sum_{j=0}^{N-1} e^{-i\delta\lambda jl} e^{ij\delta b} \psi_T(j\delta)\,\delta}_{(*)} \tag{4.40}$$

and the term $(*)$ is a discrete Fourier transform (4.21) if

$$\lambda\delta = \frac{2\pi}{N}.$$

To compute $(*)$, the user only needs to plug $e^{ij\delta b}\psi_T(j\delta)\,\delta$ in the R command *fft* of the *stats* package.

In the case of Heston and Nandi (2000), the computation of the option prices using this FFT approach can be done using the following code. The first function computes the moment generating function of the Heston and Nandi (2000) process using Proposition 3.7.1.

```
cf_hn<-function(para,vol,spec,w){

# "para" contains the model parameters in a specific order
# "vol" contains the volatility expressed in a daily manner,
# i.e  the yearly volatility should be equal to vol*(250)^(1/2)
# "spec" contains the option contract specifications:
```

```
# - time tot maturity ("spec$T")
# - strike price ("spec$K")
# - underlying asset price ("spec$S")
# - risk free rate ("spec$r")

  omega=para[1]
  beta=para[2]
  alpha=para[3]
  gamma=para[4]
  lambda=para[5]
  h=vol*vol

# Change in the parameter when moving from the historical to
# the risk neutral distribution

  gamma_star=gamma+lambda+1/2

# Terminal conditions for the A and B

A=0
  B=0

# transforms the time to maturity expressed in terms of years in terms of days

T=spec$T
   steps=round(T*250,0)
   r=spec$r/250
   K=spec$K
   S=spec$S

  for (i in 1:steps){

   A=A+w*r+B*omega-1/2*log(1-2*alpha*B)
   C=(w*w)/2-2*alpha*gamma_star*B*w+alpha*B*gamma_star*gamma_star
   D=1-2*alpha*B
   B=-1/2*w+beta*B+C/D

   }

fc=exp(w*log(S)+A+B*h)

return(fc)

}
```

And now the option pricer is obtained in the following way[21]:

```
pricer_fft<-function(para,vol,spec){

r=spec$r
alpha=2
N=2^12
delta=0.25
lambda=(2*pi)/(N*delta)
j=seq(1,N,1)
```

[21]In R, the complex number i is denoted by 1i.

```
k=seq(1,N,1)
b=(lambda*N)/2
strike=-b+(k-1)*lambda
strike=exp(strike)

res=c()

for (i in 1:N){

     wtemp=delta*(i-1)
     wtemp2=wtemp-(alpha+1)*1i
     phi=fc_hn(modele,spec,wtemp2*1i)
     phi=phi*exp(-r*T)
     phi=phi/(alpha^2+alpha-wtemp^2 + 1i*(2*alpha+1)*wtemp)
     phi=phi*exp(1i*wtemp*b)
     res=rbind(res,phi)

}

option_prices=Re(fft(res))*exp(-alpha*(-b+(k-1)*lambda))/pi

# lookup for the strikes of interest

K=spec$K
result=c()

for (i in 1:length(K)){

  index=which(strike<=K[i])
  index=index[length(index)]
  result=rbind(result,option_prices[index])

}

return(result)

}
```

Using this pricing methodology, we now compare the option pricing performances obtained with the two different approaches:

1. First, by performing what is called a "calibration" of the model, that is selecting the models' parameters so that to minimize the pricing errors for a given day. This is somewhat the ideal situation, as in such a case, parameters are forced to make the model fit the existing prices. This will provide us with a reliable benchmark: it is almost impossible that an option pricing model generates prices exactly in line with market prices, as those prices can be affected by various sources of noise, such as various liquidity effects. Of course, a model with as many parameters as the number of option prices would fit those perfectly. However it would loose its economic interpretation, not to mention the potentially very bad hedging performances that would come from such a model. By calibrating the model, we will somewhat obtain the best pricing performances achievable with a model that provides its user with a structural view of the mechanics of financial markets.
2. Then, we will assess the pricing performances of the "historical" approach, that is the approach based first on the estimation of the parameters using historical

time series of returns, and then price the options based on a change in probability measure. This approach is also historical in the sense that it is the one proposed by the seminal article of Black and Scholes (1973). The pricing errors will be of course bigger than that obtained from a "calibration" approach. Our interest lies in the comparison of those two types of performances in order to understand how different those two are. In the end, if the model provides its user with a correct specification of what is at work in financial markets, the pricing errors should not be that much different. At least, the comparison between the two scales of errors will help the reader understand the shortcomings of conditionally Gaussian models.

For both those approaches, the assessment of the quality of an option pricing model will be based on a widely accepted pricing criterion, that is the average absolute relative pricing errors. Building on the previous notations, this criterion writes as follows:

$$H(\theta) = \frac{1}{N}\sum_{i=1}^{N}\left|\frac{C(\theta, K_i, T_i, S_i, r_i) - C^M(K_i, T_i, S_i, r_i)}{C^M(K_i, T_i, S_i, r_i)}\right|, \tag{4.41}$$

where $C(\theta, K_i, T_i, S_i, r_i)$ is the option price obtained from a given model whose parameters are contained in the vector θ, for the ith observation in the dataset. Each observation is an option price for a given time to maturity T_i, a strike price K_i, a value for the underlying asset S_i and the risk-free rate r_i. The corresponding market price is $C^M(K_i, T_i, S_i, r_i)$. The lower the value of this criterion and the better is the model fit. Given its expression, the criterion is expressed as a percentage of the existing market prices and it has a minimum value of 0 % in the case of a perfect fit. In the rest of this section, we will label this criterion "AARPE" for Average Absolute Relative Pricing Errors.

Using this criterion, we first investigate the performances and results obtained when calibrating the model parameters to the existing option prices. The empirical analysis has been performed as follows:

1. First, for a given date in our sample, we gather the strikes and times to maturity for which we have available prices.
2. Then, using an optimization routine, we look for a set of parameters that minimizes $H(\theta)$ (computing the option prices using the FFT scheme presented previously) and compute the associated pricing errors for the optimal set of parameters θ^*.
3. Next, we move to the following day and start this routine again.

Hence, by doing so, we use the Heston and Nandi (2000) model in a way that is very close to what practitioners are actually doing. By focusing on a single day of option prices, the calibration will therefore help us to conclude on the minimum achievable pricing errors using a model of this sort. Two kinds of quantities are

Table 4.3 Calibrated parameters for the Heston and Nandi (2000) model

	a_0	b_1	a_1	γ
Average	6.86E−05	0.589	4.19E−07	463.3
Standard deviation	8.89E−05	0.000	1.17E−06	0.0
Quantile 2.5 %	7.55E−06	0.589	−3.00E−06	463.3
Quantile 97.5 %	3.45E−04	0.589	1.59E−06	463.3

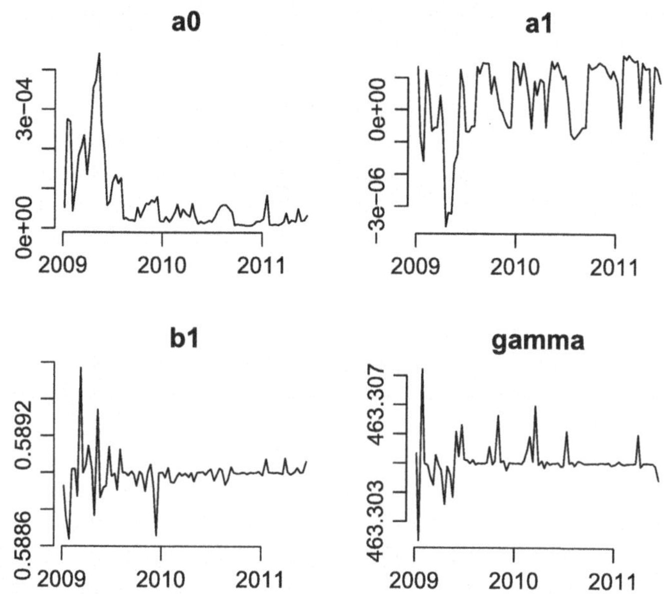

Fig. 4.8 Rolling calibration of the Heston and Nandi model over 2009–2011

worth analyzing: the variation in the calibrated parameters and a breakdown of $H(\theta)$ depending on the options' times to maturity and moneynesses. This empirical analysis is based on the S&P500 option dataset described in Sect. 4.1.

Table 4.3 and Fig. 4.8 present the values obtained when calibrating the parameters to the option prices.

We previously presented estimation results for the Heston and Nandi (2000) model under the historical distribution in Table 4.2 and in Fig. 4.7. Those two new tables present similar results under the risk neutral distribution. Several important conclusions can be reached from them:

1. Globally speaking, the scale of the parameters under both measures looks very similar, with minor differences.
2. Then, again, building an interpretation of individual parameters is difficult. It is however possible to build comparisons between groups of parameters that have an economic interpretation. Those differences are the following:

- First of all, when computing the long term volatility associated to each set of parameters, we obtain an interesting difference. In the case of historical estimates, the long term volatility is equal to 0.175 when in the risk neutral case, we obtain 0.232 that is a larger value. This is consistent with the existence of a negative volatility risk premium already mentioned.
- Next, the level of persistence is changed as well. The persistence—that is the group of parameters relating today variance with yesterday one—is in fact $b_1 + a_1\gamma^2$. This group of parameters is equal to 0.984 under the historical distribution and to 0.679 under the risk neutral one. Hence, the Heston and Nandi (2000) model diagnoses a lower persistence of volatility shocks under the risk neutral distribution when compared to the time series dynamics of returns. This is consistent with the results obtained in Barone-Adesi et al. (2008) that also report—using a different approach but still building a comparison between GARCH estimates under the historical and the risk neutral distribution—a lower degree of persistence under the risk neutral distribution.
- Finally, the leverage parameter[22] γ is found to be stronger under the risk neutral probability (463.8) than under the historical one (462.3). This is also consistent again with the skewness premium usually observed under the risk neutral probability.

3. Finally, when comparing the quantiles and the standard deviations of our estimates between Tables 4.2 and 4.3, it appears that the calibration produced a larger variation in the estimated parameters along the rolling calibration than the time series approach in the case of a_0 and a_1. The parameters b_0 and γ being rather stable, the calibration adjusts a_0 and a_1 to fit the long term value of the risk neutral volatility. Another difference can be seen in those results: the time series estimates exhibit trends that reflect the adaptation of the model to changing market conditions. The calibration does not lead to such trends: this illustrates the major difference between those approaches, from fitting daily volatility surfaces to capturing historical patterns.

Once those parameters interpretations have been cleared out, we can turn to the analysis of Table 4.4 and Fig. 4.9 that present the pricing performances of the calibrated Heston and Nandi (2000) model.

Globally speaking the pricing errors are low, ranging between 1.24 % and 34.41 % of the market price. Even though this latter percentage seems to be high, it only comes from options with a strike price higher than the current value of the underlying asset—deep out-of-the-money options. For those options, the full

[22]When under the risk neutral distribution, we refer to γ^* (see Eq. (4.16)) and to γ under the historical distribution (see Eq. (4.12)).

Table 4.4 Pricing performances (in %) obtained when calibrating the Heston and Nandi (2000) model

	Pricing performances									
Maturity/Moneyness	0.80	0.84	0.89	0.93	0.98	1.02	1.07	1.11	1.16	1.20
Maturity ≤ 0.3	1.24	1.48	2.00	3.01	6.05	8.90	10.66	11.36	18.92	34.41
$0.3 \leq$ Maturity ≤ 0.5	2.44	2.57	3.17	4.38	6.69	8.69	8.58	6.46	8.82	19.56
Maturity ≥ 0.5	3.88	3.60	3.53	4.10	4.97	5.32	5.23	4.72	5.51	16.92

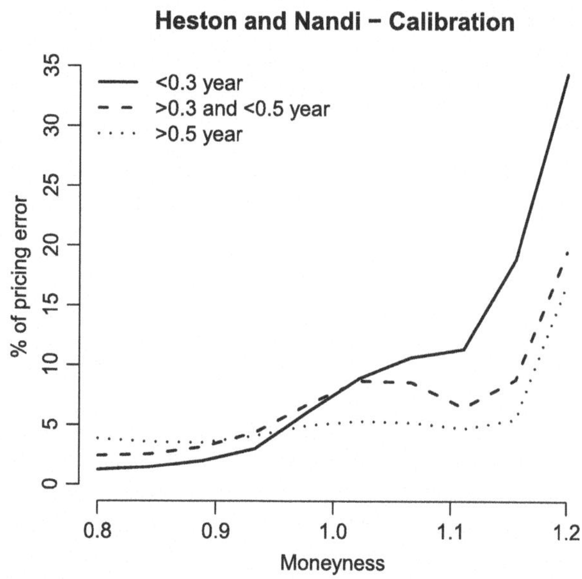

Fig. 4.9 Average absolute relative pricing errors obtained when performing a rolling calibration of the Heston and Nandi (2000) model over the January 2009–June 2011 period

price is made of its time value.[23] For such a situation, the value of the option is usually very low (less than 10$ most of the time) which makes that a pricing error would result into a few cents of mispricing. One more feature of these calibration mispricing errors is that for most moneynesses, the criterion drops when the time to maturity increases. This is the case but for the extreme parts of the moneyness grid. Hence, pricing shorter term options can prove to be trickier than pricing longer term ones. This is usually related to the sharpness of the volatility smile for shorter term options, as discussed in Bates (1988, 1996). Beyond those elements, the reader should just keep in mind that with such a perfect situation in terms of model fitting, the rough estimates for the pricing errors of a "good" option pricing

[23]The price of an European call option can be decomposed into an intrinsic value and a time value. The intrinsic value at time t is simply equal to $(S_t - K)_+$, that is what the bearer of such an option could obtain if he could exercise it immediately. The time value is defined as the difference between the current price of the option and the intrinsic value of the option. It represents how much the market values the investor's patience until the time to maturity, hence its name.

model calibrated to a single day of option prices should be between 5 and 30 %: that is the essential conclusion to be taken from these calibration results.

We now compare the previous results, obtained by calibration, to those obtained starting from the historical dynamics (4.12) estimates from which we derive the risk neutral dynamics[24] ((4.15), (4.16)) and the option prices. We add another competitor to this horse race: we also compare the pricing results obtained under the EMM $\mathbb{Q}^{qess}$ derived from the quadratic Esscher transform, as presented in Sect. 3.5. Under this new risk preference specification, and under the following constraint (see Example 3.5.1):

$$\frac{h_t^*}{h_t} = \pi > 0, \tag{4.42}$$

where h_t^* and h_t are respectively the risk neutral and the historical conditional variance[25], the associated risk neutral dynamics is described as follows:

$$Y_t = r - \frac{h_t^*}{2} + \sqrt{h_t^*}\xi_t, \quad S_0 = s \tag{4.43}$$

$$h_t^* = \pi a_0 + \pi^2 a_1 (\xi_{t-1} - (\frac{\gamma}{\pi} + \frac{\lambda_0}{\pi} + \frac{1}{2})\sqrt{h_{t-1}^*})^2 + b_1 h_{t-1}^*, \tag{4.44}$$

where the ξ_t are i.i.d $\mathcal{N}(0, 1)$, under $\mathbb{Q}^{qess}$. Comparing the results obtained from the calibration exercise to those two historical distribution-based approaches to option pricing will help the reader understand the fundamental differences in terms of performances between those various ways of pricing options using the very same model.

We present in Figs. 4.10, 4.11 and 4.12, the AARPE criterion for those three competitors for three types of times to maturity—below 3 months, between 3 and 6 months and higher than 6 months—depending on the moneyness:

Tables 4.5 and 4.6 respectively present the detailed results obtained under the Esscher and the quadratic Esscher EMM:

From these later tables, the scale of values obtained differs from the one obtained from the calibration exercise. This is however not a surprise: the calibration being sort of an ideal situation, the two historical approaches can not really match its results. Still, by gauging how far from each others those results are, the reader will understand what challenges this time series approach to option is faced with. Globally speaking, the calibration results are always the best ones. In the case of options with a time to maturity superior to 3 months, the Esscher EMM always

[24]Remind that the risk neutral dynamics ((4.15), (4.16)) is implied by the conditional Esscher transform and that, in the conditionally Gaussian case, it coincides with the one given by the extended Girsanov principle (see Sect. 3.3) and the one coming from the LRNVR in Proposition 3.2.2.

[25]In our empirical results, we use the squared value of the VIX as a measure of the risk neutral variance to determine the value of π.

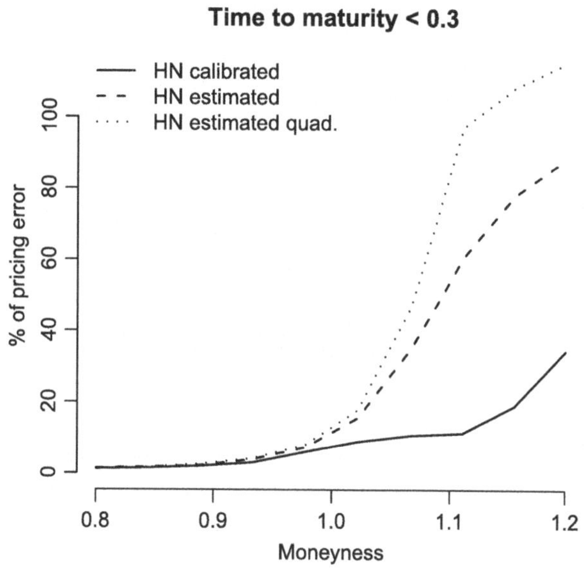

Fig. 4.10 Comparisons of the calibrated and estimated pricing performances of the Heston and Nandi (2000) model, for options with a time to maturity below 0.3 year

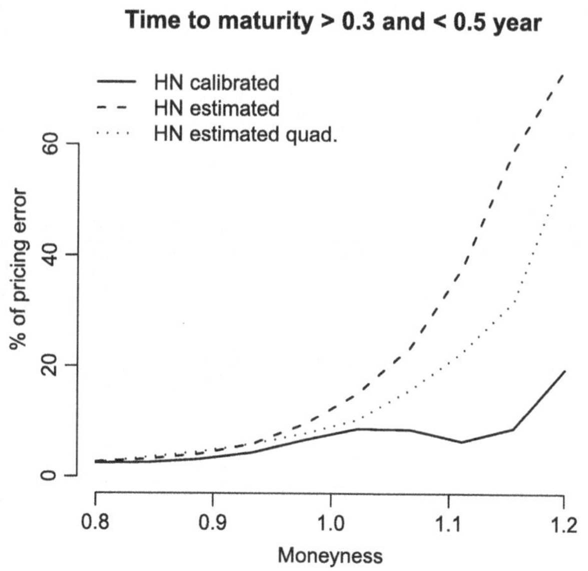

Fig. 4.11 Comparisons of the calibrated and estimated pricing performances of the Heston and Nandi (2000) model, for options with a time to maturity between 0.3 and 0.5 year

delivers the worst performances. However, for moneynesses close or below one, the pricing performances look rather similar to that of the calibration approach. For example, in the case of a time to maturity between 3 and 6 months and a moneyness of 0.98, the Esscher approach AARPE is 9.58 % when the calibration one is 6.69 %. This small difference of roughly speaking 3 % points is obtained from pricing methods with massive differences in terms of data inputs. Still, large differences exist between those two in the case of options with shorter maturities

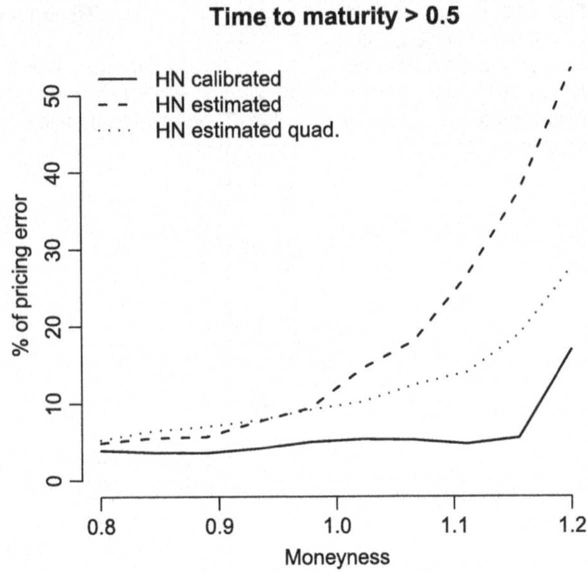

Fig. 4.12 Comparisons of the calibrated and estimated pricing performances of the Heston and Nandi (2000) model, for options with a time to maturity higher than 0.5 year

Table 4.5 Pricing performances (in %) obtained when using the time series estimates of the Heston and Nandi (2000) model and the Esscher EMM

	Pricing performances									
Maturity/Moneyness	0.80	0.84	0.89	0.93	0.98	1.02	1.07	1.11	1.16	1.20
Maturity ≤ 0.3	1.37	1.70	2.19	3.95	7.35	15.57	34.95	60.45	77.99	88.40
$0.3 \leq$ Maturity ≤ 0.5	2.66	3.20	4.12	5.96	9.58	15.12	23.44	37.99	59.47	74.35
Maturity ≥ 0.5	4.88	5.53	5.65	7.65	9.38	14.50	18.07	26.51	37.43	53.40

Table 4.6 Pricing performances (in %) obtained when using the time series estimates of the Heston and Nandi (2000) model and the quadratic Esscher EMM

	Pricing performances									
Maturity/Moneyness	0.80	0.84	0.89	0.93	0.98	1.02	1.07	1.11	1.16	1.20
Maturity ≤ 0.3	1.33	1.68	2.60	4.22	8.09	17.86	46.71	97.23	108.34	115.31
$0.3 \leq$ Maturity ≤ 0.5	2.57	3.56	4.54	5.97	7.88	10.36	15.83	22.65	31.81	56.77
Maturity ≥ 0.5	5.29	6.47	7.00	7.83	9.19	10.20	12.40	14.02	18.86	27.65

and in the case of moneyness higher than 1. As we will present it later in Table 4.11, the total AARPE for the calibration approach is 7 % when the historical one using the Esscher EMM is 24 %. The difference may seem important, but again it is not that different given the spread between the underlying dataset used to derive both risk neutral distributions.

A first way to improve the pricing performances without giving up on the Heston and Nandi (2000) mechanics is to use more complex risk preferences. The previously mentioned quadratic Esscher transform (see Sect. 3.5) provides its user with a way to do so, while maintaining a closed-form expression for the characteristic function of the process. The pricing results show that using such a stochastic discount factor leads to a decrease in the global AARPE: for the Esscher EMM the AARPE is equal to 24 % and in the case of the quadratic Esscher one this figure shrinks to 22 %. The improvement is not massive, but when investigating the breakdown of the AARPE by time to maturity, the improvement seems more impressive. Figures 4.10, 4.11 and 4.12 provide such comparisons: in the case of options with a time to maturity larger than 3 months and especially for moneynesses larger than 1, the quadratic Esscher EMM leads to a significant improvement in reducing the pricing errors. For example, in the case of a moneyness equal to 1.16 and a time to maturity between 3 and 6 months, the Esscher EMM generates an AARPE equal to 59.47 % when the quadratic one's is equal to 31.81 %. The only exception is obtained for shorter term options. For a such a case, improving the specification of risk preferences does not improve the pricing performances from our dataset perspective. The conditional distribution used for the Heston and Nandi (2000) model being Gaussian, it stands a limited chance to be able to capture the behavior of shorter term options that can only be generated by fat-tailed processes. In the next subsection, we investigate the pricing performances obtained from those types of processes using the distributions introduced in Sect. 2.4.

4.3 Empirical Performances of Heavy Tailed Models

The last section put the emphasize on two key points: first, when performing a calibration using the benchmark model of the industry—the Heston (1993) model—the AARPE ranges from 1 to 30 %. Second, when starting from the historical analysis of the very same model, the performance ended up to be disappointing when compared to what calibration can come up with. One of the key reasons can be found in a careful analysis of the evolution of the parameters estimated using both methods: when the historical method captures trends related to changing financial conditions, the calibration systematically looks for the best fit, making parameters change from 1 day to another with no fundamental economic reason. Still, what matters to a quantitative analysis is the pricing performance, to which the hedging performance will be related: in that respect, calibration is king. The historical approach can still make a difference there, provided that the underlying time series model is able to mimic both the time varying volatility observed in financial markets and the fat tails of returns. This last section will offer the reader a tour around adequate time series models, emphasizing what values for the parameters can be expected and what option pricing performances can be achieved with them.

4.3.1 Estimation Strategies

In this final section, we investigate the pricing performances of conditionally non-Gaussian models: by doing so, we give up on the closed-form expression for the characteristic function, hoping that the increase in the goodness of fit of the returns' distribution will be worth the increase in the time necessary to compute the price of an option. Under this specification, the returns' dynamics is the following:

$$Y_t = \log\left(\frac{S_t}{S_{t-1}}\right) = r + \lambda_0\sqrt{h_t} - \frac{h_t}{2} + \sqrt{h_t}z_t \tag{4.45}$$

$$h_t = F(z_{t-1}, h_{t-1}), \tag{4.46}$$

where z_t are i.i.d. random variables following an asymmetric distribution with fat tails. The variance dynamics that we use are the EGARCH and the APARCH models introduced in Sect. 2.3.[26] Those two models are both able to accommodate volatility clustering effects and an asymmetric reaction of variance to past returns, depending on their sign. The EGARCH dynamics is given by

$$\log h_t = a_0 + a_1(|z_{t-1}| - \gamma z_{t-1}) + b_1 \log h_{t-1} \tag{4.47}$$

and the APARCH one is the following:

$$h_t^{\frac{\Delta}{2}} = a_0 + a_1(|\sqrt{h_{t-1}}z_{t-1}| - \gamma\sqrt{h_{t-1}}z_{t-1})^{\Delta} + b_1 h_{t-1}^{\frac{\Delta}{2}}. \tag{4.48}$$

In our empirical study, z_t follows three different distributions that all have fat tails:

- The Generalized Hyperbolic distribution (see Sect. 2.4.1):

$$z_t \hookrightarrow GH(\lambda, \alpha, \beta, \delta, \mu). \tag{4.49}$$

- A mixture of two Gaussian distributions (see Sect. 2.4.2):

$$z_t \hookrightarrow MN(\phi, \mu_1, \mu_2, \sigma_1, \sigma_2) \tag{4.50}$$

- A jump component (see Sect. 3.4.2):

$$z_t = z_t^0 + \sum_{i=1}^{N_t} z_t^i, \tag{4.51}$$

where $(z_t^0, N_t, z_t^i)_{i\in\mathbb{N}^*, t\in\{1,\ldots,T\}}$ are independent random variables such that z_t^0 follows a $\mathcal{N}(0,1)$, N_t follows a $\mathcal{P}(\lambda)$ and z_t^i follows a $\mathcal{N}(\mu,\sigma^2)$.

Mixing these conditional distributions with the previous variance dynamics makes it possible to capture both leverage effects and fat tails in the conditional

[26] We do not study the GJR specification in this section because we have seen in Table 2.4 that it is a special case of the APARCH model.

distribution: they are sources of asymmetries and leptokurticity in both the variance equations and the conditional distributions.[27]

Compared to the Heston and Nandi (2000) model that is based on 5 parameters, including the risk premium, we have now between 7 (for the EGARCH-Jump) and 10 (for the APARCH-GH and APARCH-MN) parameters to fit the returns' distribution. In this respect—and knowing that any additional parameter is an additional degree of freedom to fit the sample's distribution—there is no need to test for the superiority of those models over the Heston and Nandi (2000) one in terms of goodness of fit under the historical distribution: they do provide a better fit. This does not say much about their superiority in terms of capturing the mechanism of financial markets, as it is possible that those models actually over-fit the sample, that is that they include too many parameters for the data that we are trying to bring into equations. Without using complex econometric tests and so that to fit the purpose of this book, there are two ways of checking that those models do not over-fit the data. The first one consists in investigating a rolling estimation, as we did previously for the HN model. The second one will consist in testing the ability of those models to deliver accurate option prices.

The first issue will thus be to estimate the parameters by maximum likelihood. Compared to the HN case, we need to deal with an additional difficulty: the Heston and Nandi (2000) model incorporates parameters that have individually a well-defined impact on the returns' distribution. For example, a_0 somewhat will guide the average level of variance, when γ handles leverage effects and a_1 deals with the thickness of the tails. In our mixing cases, we have a similar well-defined interpretation for each of the parameters of the conditional distributions and of the variance when considered separately. Now, when bringing them together in a single time series model, parameters from both parts will have a very similar impact on the returns' distribution, without being fully comparable from an economic perspective. For example, in the EGARCH-GH case, the parameter γ from the EGARCH structure and the parameter β from the GH distribution will impact the returns' asymmetry. However, even though they impact the distribution the same way, they do so based on a different rationale: in the EGARCH case, this asymmetry comes from the past returns to volatility feedback channel that is usually referred to as "leverage effect". In the GH case, β captures the likely sign of extreme events: when it is negative, extreme events will have an increased tendency to be of a negative sign. Hence, in the S&P500 case, we can expect both those effects to co-exist, as both leverage effects and extreme events are widely accepted stylized facts in the case of equity indices. Given that our estimation strategy will be based on the returns' distribution that is used in the likelihood function that needs to be

[27] As introduced in Chaps. 2 and 3, we will refer to these mix of variance dynamics and distributions using EGARCH and APARCH for the two volatility structures and GH for the Generalized Hyperbolic distribution, MN for the Mixture of Gaussian distributions and "Jumps" for the jump component. The "EGARCH-Jump" model will therefore be the mixture of an EGARCH variance dynamics with the jump component as a conditional distribution.

maximized to obtain estimates, we stand a good chance that a global maximization of the likelihood function over the full set of parameters will deliver poor results. Beyond those mixing effects, the likelihood function stands a good chance to have local maxima that will attract the numerical optimizer. To circumvent this issue, we rely on the REC estimation strategy that has been presented in Sect. 2.6.3: this estimation strategy proved to be able to provide reliable small-sample estimates and to disentangle what comes from the conditional distribution and what comes from the variance dynamics.[28] We do not provide in this book the full code that performs such a recursive estimation scheme given that it simply boils down to recursively optimizing the full likelihood over subsets of parameters. We only provide the code necessary to compute the log-likelihood for the models that we investigate. In each case, this log-likelihood function is computed as follows (see Sect. 2.6):

$$L_T(y_1, \ldots, y_T \mid \theta) = \sum_{i=1}^{T} log\left(d^{\mathbb{P}}_{Y_i|\mathcal{F}_{i-1}}(y_i)\right), \tag{4.52}$$

where $d^{\mathbb{P}}_{Y_i|\mathcal{F}_{i-1}}(y_i)$ is the conditional density associated to a given model at the point y_i, the ith observation.

This first piece of code computes the variance given the past variance and the past innovation:

```
variance<-function(type_vol,para_vol,vol_old,innovation)
{

# This function computes the volatility for both variance structures
# type_vol is string containing the name of the variance structure
# para_vol is a vector containing the variance parameters
# vol_old contains the previous volatility
# innovation contains the previous z_t

  if(type_vol=="EGARCH")
  {
  a0=para_vol[1]
  a1=para_vol[2]
  gamma=para_vol[3]
  b1=para_vol[4]
  vol=a0+a1*(abs(innovation)-gamma*innovation)+b1*log(vol_old^2)
  vol=exp(vol)
  vol=sqrt(vol)
  }

  if(type_vol=="APARCH")
  {
  w=para_vol[1]
  a=para_vol[2]
  gamma=para_vol[3]
  b=para_vol[4]
```

[28] An interested reader will read with interest Chorro et al. (2014).

```
  Delta=para_vol[5]
  vol=w+a*(abs(vol_old*innovation)-gamma*vol_old*innovation)^(Delta)
   +b*vol_old^(Delta/2)
  vol=vol^(2/delta)
  vol=sqrt(vol)
  }

return(vol)

}
```

The next piece of code computes the conditional density for the three distributions used here:

```
density<-function(type_distribution,para_distribution,vol,x,r,m)
{

# This function computes the likelihood associated to each conditional
# distribution
# type_distribution contains a string that names the conditional
# distribution
# para_distribution is a vector of parameters
# vol is the conditional volatility
# x contains the asset's return
# r is the risk-free rate
# m is the risk premium


if (type_distribution=="GH")
{
alpha=para_distribution[1]
beta=para_distribution[2]
delta=para_distribution[3]
mu=para_distribution[4]
lambda=para_distribution[5]
result=dgh(x,alpha/vol,beta/vol,delta*vol,r+m+mu*vol,lambda)
}

if (type_distribution=="MN")
{
phi=para_distribution[1]
mu1=para_distribution[2]
sigma1=para_distribution[3]
mu2=para_distribution[4]
sigma2=para_distribution[5]
result=phi*dnorm(x,r+m+vol*mu1,sigma1*vol)
      +(1-phi)*dnorm(x,vol*mu2+r+m,sigma2*vol)
}

if (type_distribution=="GAUSSIAN")
{
  resultat=dnorm(x,m+r,vol)
}

if (type_distribution=="Jumps")
{
```

```
lambda=para_distribution[1]
mu=para_distribution[2]
sigma=para_distribution[3]

m_star=m-(vol*lambda*mu)/sqrt(1+lambda*(mu^2+sigma^2))
h_star=(vol^2)/(1+lambda*(mu^2+sigma^2))

k = 1:100
  temp1=((lambda^k)*exp(-lambda))/factorial(k)
  vol_temp=(1+k*sigma^2)*h_star
  vol_temp=sqrt(vol_temp)
  temp2=dnorm(x,r+m_star+sqrt(h_star)*k*mu,vol_temp)
  result=sum(temp1*temp2)
}

return(result)
}
```

Finally, the next function computes the log-likelihood associated to a given sample of returns, using the previous two functions:

```
likelihood_total<-function(para,type_distribution,type_vol,rend,r)
{

# This function computes the log-likelihood associated to the mixed
# models
# para contains the parameters in the following order: the risk premium,
# the distribution parameters and finally the variance parameters
# type_distribution contains a string that names the distribution
# type_vol does the same for volatility
# rend contains a vector of returns
# r contains the corresponding risk-free rates.

lambda_0=para[1]

if (type_distribution=="GAUSSIAN"){
para_vol=para[-1]
para_distribution=c()
}

if (type_distribution=="MN"){
para_distribution=para[2:6]
para_vol=para[7:length(para)]
}

if (type_distribution=="GH"){
para_distribution=para[2:6]
para_vol=para[7:length(para)]
}

if (type_distribution=="Jumps"){
para_distribution=para[2:4]
para_vol=para[5:length(para)]
}

# Calcul de la vraisemblance
```

```
vol=sd(rend)
f<-0
for (i in 1:length(rend))
{
  m=lambda_0*vol-(vol^2)/2
  temp=density(type_distribution,para_distribution,vol,rend[i],r[i],m)
  f<-f+log(temp)
  innovation=(rend[i]-r[i]-m)/vol
  vol=variance(type_vol,para_vol,vol,innovation,mean_distribution)
}

return(-f/length(rend))
}
```

Now, using the previous code to estimate the parameters using the recursive estimation methodology, we obtained Table 4.7 that presents the estimates obtained with our sample of returns. Two things are worth mentioning in this table. First of all, it provides the reader with potential values that can be used as starting values for the optimization of the likelihood function: even though we use a specific sample—estimated parameters over a different sample would be different—the values that we obtain will still provide the reader with rough estimates that are essential to start any numerical work around those models. For example, b_1 from the variance equations is roughly speaking higher than 0.9, as in the Heston and Nandi (2000) case under the historical distribution. The a_1 parameter looks as well very similar in our estimations, and close to 0.1. A similar comment can be made in the case of the leverage parameter γ, that evolves around 0.8. Next, the GH parameters estimated using the EGARCH and the APARCH variance structures are close to each others: β is found to be negative, underlining the fact that residuals still have a negative skewness that is not dealt with by the leverage effect of the volatility structure. For both cases, α is roughly speaking smaller than twice the absolute value of β. Still, it is large enough for the distribution of the residuals to have fat tails. In the mixture of Gaussian distributions case, the shape of the two mixtures is similar across our estimations: it is made of a distribution with a high variance and strongly negative expectation with a low probability of occurrence—that is by essence a distribution that accounts for extreme events—and a distribution with a variance below 1 and an expectation very close to 0. The mixture of the two makes it possible to thus generate a conditional distribution with a negative skewness and a kurtosis larger than 3, as in the GH case. Finally, the estimated parameters in the jump case are also consistent with fat tails: the estimated jump components are made of a high number of jumps with an expectation very close to zero and small variance. For each of those conditional distributions, the estimated expectations and variances are respectively very close to 0 and to 1. This constraint is essential to enforce the fact that h_t remains the conditional variance of the returns. The second conclusion from this table is that for most of the parameters, we observe a low variability of the estimated values over the dates for which we performed our rolling estimation, as it can be seen from the two quantiles that are provided in Table 4.7. This somewhat means that despite the

Table 4.7 Estimated parameters across time series models

		λ_0	a_0	a_1	γ	b_1	α	β	δ	μ	λ	
EGARCH-GH	Average	0.080	−0.232	0.132	0.819	0.986	0.856	−0.561	2.786	0.379	−5.614	
	Standard deviation	0.055	0.012	0.007	0.055	0.001	0.132	0.085	0.195	0.084	0.577	
	Quantile 2.5 %	−0.012	−0.254	0.119	0.711	0.984	0.641	−0.701	2.423	0.226	−6.895	
	Quantile 97.5 %	0.174	−0.212	0.146	0.932	0.987	1.141	−0.427	3.089	0.532	−4.653	
		λ_0	a_0	a_1	γ	b_1	ϕ	μ_1	σ_1	μ_2	σ_2	
EGARCH-MN	Average	0.068	−0.227	0.134	0.865	0.986	0.055	−1.103	1.405	0.000	0.814	
	Standard deviation	0.060	0.010	0.008	0.073	0.001	0.011	0.096	0.117	0.058	0.049	
	Quantile 2.5 %	−0.033	−0.248	0.121	0.735	0.983	0.037	−1.309	1.231	−0.093	0.734	
	Quantile 97.5 %	0.160	−0.209	0.150	0.969	0.987	0.073	−0.956	1.626	0.098	0.908	
		λ_0	a_0	a_1	γ	b_1	λ	μ	σ			
EGARCH-Jumps	Average	0.018	−0.235	0.116	0.841	0.984	15.702	0.000	0.017			
	Standard deviation	0.025	0.009	0.003	0.027	0.001	1.798	0.002	0.035			
	Quantile 2.5 %	−0.073	−0.253	0.110	0.805	0.983	10.166	−0.002	0.000			
	Quantile 97.5 %	0.042	−0.223	0.120	0.907	0.986	18.787	0.007	0.151			
		λ_0	a_0	a_1	γ	b_1	Δ	α	β	δ	μ	λ

APARCH-GH	Average	0.034	3.80E−05	0.110	0.800	0.909	1.350	1.396	−0.880	2.581	0.497	−6.181
	Standard deviation	0.028	1.70E−06	1.08E−04	3.53E−05	2.41E−04	0.026	0.209	0.115	0.107	0.064	0.439
	Quantile 2.5 %	−0.034	3.55E−05	0.110	0.800	0.909	1.301	1.075	−1.101	2.378	0.392	−6.822
	Quantile 97.5 %	0.073	4.06E−05	0.110	0.800	0.910	1.386	1.764	−0.706	2.764	0.604	−5.212
		λ_0	a_0	a_1	γ	b_1	Δ	ϕ	μ_1	σ_1	μ_2	σ_2
APARCH-MN	Average	0.032	3.87E−05	0.110	0.800	0.909	1.354	0.052	−1.011	1.299	0.028	0.736
	Standard deviation	0.013	1.35E−06	2.91E−05	9.56E−06	3.99E−04	0.016	0.016	0.138	0.047	0.016	0.006
	Quantile 2.5 %	0.003	3.72E−05	0.110	0.800	0.909	1.324	0.033	−1.224	1.210	−0.004	0.722
	Quantile 97.5 %	0.056	4.10E−05	0.110	0.800	0.911	1.383	0.092	−0.753	1.373	0.057	0.745
		λ_0	a_0	a_1	γ	b_1	Δ	λ	μ	σ		
APARCH-Jumps	Average	−0.176	4.31E−05	0.109	0.800	0.909	1.314	12.332	0.015	0.011		
	Standard deviation	0.032	1.82E−06	5.34E−04	8.66E−05	2.08E−04	0.014	2.755	0.004	0.004		
	Quantile 2.5 %	−0.263	4.03E−05	0.107	0.800	0.909	1.290	8.974	0.009	0.003		
	Quantile 97.5 %	−0.116	4.67E−05	0.109	0.801	0.910	1.350	19.206	0.021	0.018		

Parameters estimates have been obtained for each working day from January 7, 2009 to June 29, 2011 using the REC estimation strategy and 4,000 log-returns

Table 4.8 Density tests testing the accuracy of different estimation strategies for considered GARCH models

Models	ML vs. QML	REC vs. QML	REC vs. ML
EGARCH-GH	2.956	3.922	1.975
APARCH-GH	2.847	3.018	1.934
GJR-GH	3.015	3.119	2.002
EGARCH-MN	3.132	3.052	2.027
APARCH-MN	4.820	4.823	2.806
GJR-MN	4.797	5.811	2.854
EGARCH-Jump	3.807	4.804	2.845
APARCH-Jump	3 .792	3.850	2.816
GJR-Jump	4.780	4.706	2.759

This table presents the results of the Amisano and Giacomini (2007)'s test comparing, in sample, the different estimation strategies for considered GARCH models. The data set used starts from January 3, 1990 to June 29, 2011. ML stands for maximum likelihood, QML stands for Quasi Maximum Likelihood and REC stands for Recursive Likelihood. The test reads as follows: in the EGARCH-GH case, the column "ML vs. QML" uses the estimated parameters by ML as model 1 and the parameters estimated by QML as model 2. The test statistics (4.53) value is 2.956: this value being outside the $[-1.96 : 1.96]$ 5 % interval confidence, the null hypothesis that both models are equivalent is rejected. The positivity of this statistics indicates that model 1 is favored over model 2

fact that those models use more parameters than Heston and Nandi (2000)'s model, those parameters do not seem to be used to over-fit the sample. This provides us with a first positive message when it comes to the interest of using such time series models for option pricing purposes.

Before turning to the option pricing performances of the non-Gaussian models presented before, Tables 4.8 and 4.9 and Figs. 4.13 and 4.14 illustrate the importance of the estimation method used. The previous discussion regarding the mixing impact of parameters with similar impacts on the shape of the distribution of returns could appear as theoretical. Table 4.8 compares the quality of the fit obtained for one given model across the three estimation strategies described in Sect. 2.6: the maximum likelihood, the quasi maximum likelihood and the recursive estimation. To compare the estimation methods we propose to use the joint density of the sample as a score. The test we use can be seen as an in-sample version of the test proposed for density forecast in Amisano and Giacomini (2007) as presented in Vuong (1989).[29]
For each model, the recursive estimation method yields estimates whose goodness of fit is statistically superior to that of the estimates obtained with the other two methods. This effect is obtained for every model under our scope.

[29] Say we deal with a time series model for the log-returns whose estimated conditional density at time t is $f(Y_t|\underline{Y_{t-1}}, \hat{\theta}_1)$, where $\underline{Y_{t-1}} = (Y_1, \ldots, Y_{t-1})$ and $\hat{\theta}_1$ is the vector of parameters describing the shape of this conditional distribution and the volatility structure estimated from methodology 1. We compare this estimation method to another one defined by the conditional density $f(Y_t|\underline{Y_{t-1}}, \hat{\theta}_2)$, with $\hat{\theta}_2$ being the estimated parameters obtained from this second method.

Table 4.9 Density tests testing model domination

Models	EGARCH-GH vs. APARCH-GH	EGARCH-GH vs. EGARCH-MN	EGARCH-GH vs. APARCH-MN	APARCH-GH vs. EGARCH-MN	APARCH-GH vs. APARCH-MN	EGARCH-MN vs. APARCH-MN	EGARCH-GH vs. GJR-GH	EGARCH-GH vs. GJR-MN
REC	0.847	1.018	0.934	0.979	0.874	1.028	0.827	1.057
ML	0.132	0.052	−0.027	0.042	0.092	0.064	0.195	0.159
QML	2.797	2.811	2.854	1.892	2.799	2.861	2.768	2.781
Models	EGARCH-GH vs. APARCH-Jump	EGARCH-GH vs. EGARCH-Jump	EGARCH-MN vs. EGARCH-Jump	EGARCH-MN vs. APARCH-Jump	EGARCH-MN vs. GJR-GH	EGARCH-MN vs. GJR-MN	EGARCH-MN vs. GJR-Jump	EGARCH-Jump vs. APARCH-GH
REC	0.792	0.850	0.816	0.787	0.687	0.905	0.766	0.920
ML	0.601	0.828	0.742	0.847	0.781	0.795	0.583	0.827
QML	2.084	2.904	2.848	0.788	2.137	2.033	2.086	2.056
Models	EGARCH-Jump vs. APARCH-Jump	EGARCH-Jump vs. GJR-GH	EGARCH-Jump vs. GJR-MN	EGARCH-Jump vs. GJR-Jump	APARCH-MN vs. APARCH-Jump	APARCH-MN vs. GJR-GH	APARCH-MN vs. GJR-MN	APARCH-MN vs. GJR-Jump
REC	1.084	1.087	1.109	1.271	1.202	1.070	1.046	1.116
ML	1.929	1.803	1.810	1.818	1.910	1.848	1.955	1.851
QML	2.932	2.763	1.730	2.579	1.912	1.941	1.983	1.955
Models	APARCH-Jump vs. GJR-MN	APARCH-Jump vs. GJR-Jump	GJR-GH vs. GJR-MN	GJR-GH vs. GJR-Jump	GJR-MN vs. GJR-Jump	APARCH-GH vs. APARCH-Jump	APARCH-GH vs. GJR-GH	APARCH-GH vs. GJR-MN
REC	1.859	1.737	1.691	1.739	2.000	1.754	1.791	1.786
ML	1.985	1.899	1.923	1.030	1.888	1.960	1.935	1.021
QML	2.875	2.844	1.915	2.884	2.815	2.862	2.760	2.822
Models	EGARCH-GH vs. GJR-Jump	EGARCH-Jump vs. APARCH-MN	APARCH-Jump vs. GJR-GH	APARCH-GH vs. GJR-Jump				
REC	1.090	0.935	1.089	1.666				
ML	0.097	0.867	1.826	1.057				
QML	1.911	1.890	0.806	2.853				

This table presents the results of the Amisano and Giacomini (2007)'s test comparing, in sample, different GARCH models. The data set used starts from January 3, 1990 to June 29, 2011. ML stands for maximum likelihood, QML stands for Quasi Maximum Likelihood and REC stands for Recursive Likelihood. The test reads as follows: using the REC estimation approach, when the EGARCH-GH is model 1 and the APARCH-GH is model 2, the test statistics (4.53) value is 0.847: this value being inside the [−1.96 : 1.96] 5 % interval confidence, the null hypothesis that both models are equivalent is not rejected

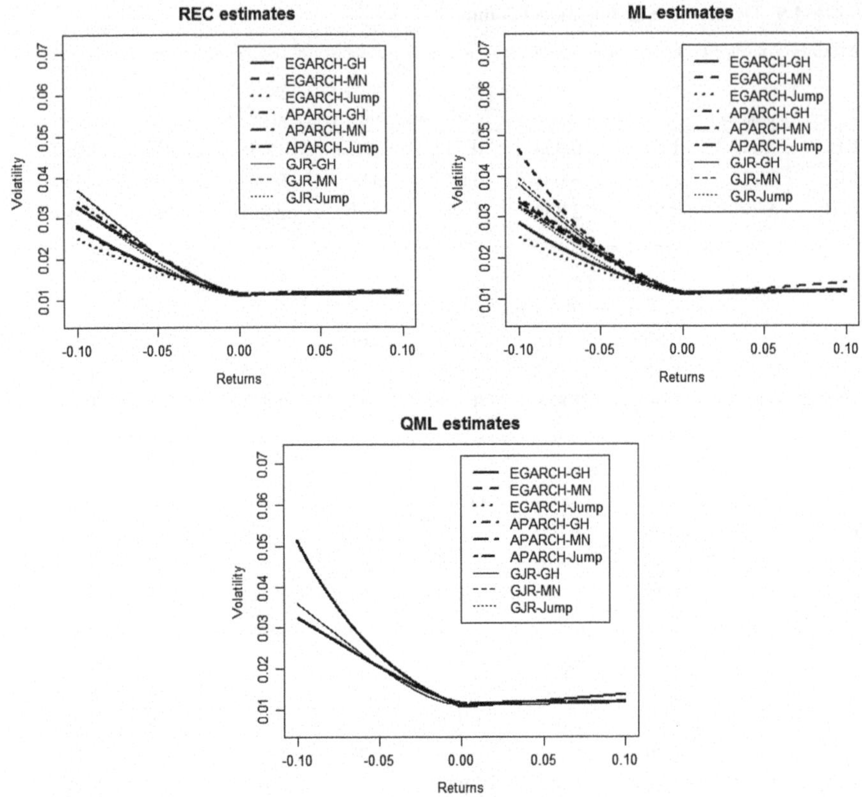

Fig. 4.13 Volatility News Impact curves. The data set used starts from January 3, 1990 to June 29, 2011. The figure presents the News Impact curve when the initial level of volatility is set to the average of the sample at hand. The parameters used are the parameters estimated with the three maximum likelihood approaches. The *top-left figure* is for the REC estimates, the *top-right* is for the ML estimates and the *bottom figure* is for the QML estimates

What is more, when investigating the news impact curves from Fig. 4.13 obtained from each estimation scheme as well as the conditional distributions as displayed in Fig. 4.14, the shapes obtained with each model are more similar in the recursive

---

The null hypothesis of the test is "methods 1 and 2 provide a similar fit of the log-return's conditional distribution using the same underlying model". The corresponding test statistic is then:

$$t_{1,2} = \frac{1}{n}\sum_{t=1}^{n}\left(\log f(Y_t|\underline{Y_{t-1}}, \hat{\theta}_1) - \log f(Y_t|\underline{Y_{t-1}}, \hat{\theta}_2)\right)$$

where n is the total number of observations available. Under the null hypothesis

$$\frac{t_{1,2}}{\hat{\sigma}_n}\sqrt{n} \underset{n\to+\infty}{\rightarrow} \mathcal{N}(0, 1), \qquad (4.53)$$

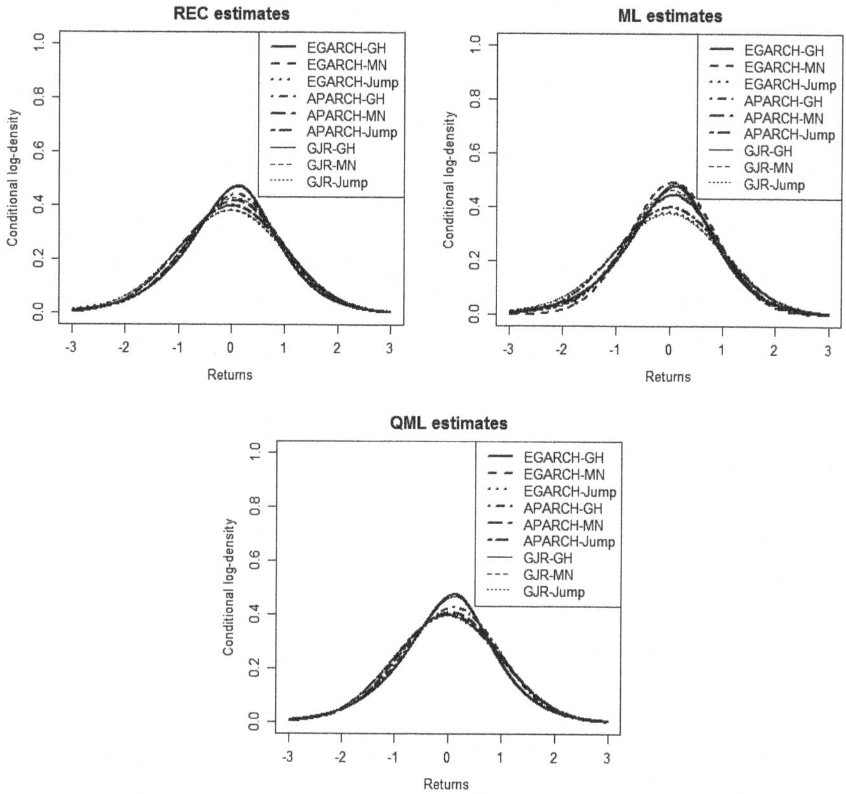

Fig. 4.14 Estimated conditional log-densities. The data set used starts from January 3, 1990 to June 29, 2011. The figure presents the estimated conditional densities. The parameters used are the parameters estimated with the three maximum likelihood approaches. The *top-left figure* is for the REC estimates, the *top-right* is for the ML estimates and the *bottom figure* is for the QML estimates

estimation case: intuitively, each of the models have the potential to provide a similar goodness-of-fit as they all include time varying volatility, leverage effect, an asymmetry in the conditional distribution as well as fat tails. Therefore, when properly estimated, their news impact curves and their conditional distributions should look somewhat similar. From the figures, the maximum likelihood and the quasi-maximum likelihood estimates yield news impact curves with a higher dispersion than those obtained with the recursive estimation strategy. Finally, Table 4.9 provides a comparison between models using the three sets of estimates, based on the Amisano and Giacomini (2007)'s test: for each pair of models, we test

where $\hat{\sigma}_n$ is a properly selected estimator for the statistic volatility. Here, as proposed in Amisano and Giacomini (2007), we use a Newey-West estimator, with a lag empirically retained to be large (around 25).

whether one is favored to the other, based on each set of estimates. When using the estimates obtained with maximum likelihood and quasi-maximum likelihood, various models are favored over others. Now, when using recursive estimates, all models are found to be statistically equivalent. Hence, the choice of the estimation method is indeed of a high importance, as it can lead to a wrong model selection. From Table 4.9, each of the non-Gaussian models are actually standing equivalent chances to deliver appropriate option prices, as they all have the key requirements necessary to mimic the historical behavior of option prices. Still, by mixing them with pricing kernels, the outcome remains uncertain: when changing of probability measure, the innovations will be changed and will impact the dynamics of volatility and higher order moments in a complex way. This is what we are now going to investigate.

4.3.2 Pricing Performances

Once the models' parameters have been estimated, we are now ready to obtain the arbitrage-free price of European options (see Definition 3.2.2) using the stochastic discount factor approach described in Sect. 3.2.2.

Even if the conditional characteristic function of each of these non-Gaussian models can be computed between time t and time $t+1$, it cannot be computed between time t and time T, where $T > t+1$, in a closed-form expression.[30] This is why we need to rely on Monte Carlo methods (see the Mathematical Appendix) to obtain the option prices associated to these models. This type of experiment is time consuming. However, we do not need to perform a calibration: each option price is only computed once, resulting in a limited time consumption in our approach.

Throughout this Monte Carlo procedure, we need to perform changes in the historical parameters in order to sample returns under a risk neutral probability, as detailed in Chap. 3. In this empirical study, we will restrict ourselves to the use of the conditional Esscher transform (see Sect. 3.4) to obtain an equivalent martingale measure well-adapted to the non-Gaussian GARCH framework.[31] Those changes result from the computation of the conditional moment generating function associated with each process and the solving of the pricing system specified in equation (3.47). The following code shows how to compute those moment generating functions:

```
chara_fun<-function(u,type_distribution,para_distribution,vol,r,m)
{

# This function computes the characteristic functions associated to
```

[30] Thus, the Carr and Madan (1999) methodology mentioned in Sect. 3.7 cannot be used in practice to price European call options.

[31] The poor pricing performances of the extended Girsanov principle approach (see Sect. 3.3) are now well-documented in the literature especially for long maturity options as remarked in Badescu and Kulperger (2008) and Badescu et al. (2011). This leads us to focus on the Esscher dynamics.

```
# each of the time series models

# u is the point for which this CF is computed
# type_distribution is the type of distribution used
# para_distribution contains the parameters driving the distribution
# vol contains the current volatility
# r contains the risk free rate
# m contains the risk premium

if (type_distribution=="GH")
{
alpha=para_distribution[1]
beta=para_distribution[2]
delta=para_distribution[3]
mu=para_distribution[4]
lambda=para_distribution[5]

A=u*(mu*vol+r+m)
A=exp(A)

B=besselK(delta*sqrt(alpha^2-(beta+u*vol)^2),lambda)/
besselK(delta*sqrt(alpha^2-beta^2),lambda)

C=(alpha^2-beta^2)/(alpha^2-(beta+u*vol)^2)
C=C^(lambda/2)

result=A*B*C
}

if (type_distribution=="MN")
{
phi=para_distribution[1]
mu1=para_distribution[2]
sigma1=para_distribution[3]
mu2=para_distribution[4]
sigma2=para_distribution[5]

m1=(r+m+vol*mu1)*u
s1=(vol*sigma1*u)^2
s1=s1/2

m2=(r+m+vol*mu2)*u
s2=(vol*sigma2*u)^2
s2=s2/2

result=phi*exp(m1+s1)+(1-phi)*exp(m2+s2)
}

if (type_distribution=="Jumps")
{
lambda=para_distribution[1]
mu=para_distribution[2]
sigma=para_distribution[3]

m_star=m-(vol*lambda*mu)/sqrt(1+lambda*(mu^2+sigma^2))
h_star=(vol^2)/(1+lambda*(mu^2+sigma^2))

A=(r+m_star)*u
B=(u^2)*h_star/2
C1=mu*sqrt(h_star)*u+(h_star*(u^2)*(sigma^2))/2
C=lambda*(exp(C1)-1)

D=A+B+C
```

```
result=exp(D)
}

return(result)
}
```

This next piece of code solves the pricing equation (3.47) to obtain the value of the Esscher parameter θ_t^q at each step of the sampling process:

$$\frac{\mathbb{G}^{\mathbb{P}}_{Y_t|\mathcal{F}_{t-1}}(1+\theta_t^q)}{\mathbb{G}^{\mathbb{P}}_{Y_t|\mathcal{F}_{t-1}}(\theta_t^q)} = e^r, \tag{4.54}$$

where $\mathbb{G}^{\mathbb{P}}_{Y_t|\mathcal{F}_{t-1}}(u)$ is the conditional moment generating function, under the historical probability $\mathbb{P}$, of Y_t given $\mathcal{F}_{t-1}$ computed at the point u.[32] The code does not look for the zero of this equation, but equivalently minimizes the following expression instead:

$$\left(\frac{\mathbb{G}^{\mathbb{P}}_{Y_t|\mathcal{F}_{t-1}}(1+\theta_t^q)}{\mathbb{G}^{\mathbb{P}}_{Y_t|\mathcal{F}_{t-1}}(\theta_t^q)} - e^r\right)^2. \tag{4.55}$$

This is why it is structured into two different pieces: the first one computes the value of the objective function for a given value of θ_t^q and the second one minimizes this expression by first looking for intervals over which θ_t^q is defined.

```
to_solve_theta<-function(theta,type_distribution,para_distribution,
vol,r,m){
# This function computes the objective function to obtain theta
# As we are looking for the zero of an expression, we minimize its
# quadratic value instead
G1=chara_fun(1+theta,type_distribution,para_distribution,vol,r,m)
G2=chara_fun(theta,type_distribution,para_distribution,vol,r,m)
res=(G1/G2-exp(r))^2
return(res)
}


solve_theta<-function(type_distribution,para_distribution,vol,r,m){
# This function solves the characteristic function based problem, using
# the  "to_solve_theta" function

# Theta typically lies between -10 and 10
theta_sup=seq(-10,10,length=1000)
temp=to_solve_theta(theta_sup,type_distribution,para_distribution,
vol,r,m)

# As it is not defined for all values, we check which value does not
# match a "NA"
index=which(is.na(temp)==FALSE)
mini=theta_sup[index[1]]
```

[32]Explicit expressions of the conditional moment generating functions for the distributions considered in this empirical part are given in Sect. 3.4.2.

```
maxi=theta_sup[index[length(index)]]

# Then we use the "optimize" function to find the minimum of the
# function over its definition set
result=optimize(to_solve_theta,c(mini,maxi),
type_distribution,para_distribution,vol,r,m)$min
return(result)
}
```

The last step to building this Monte Carlo option pricer will consist in adding to this set of functions a function that samples the risk neutral dynamics described in Sect. 3.4.2. This is what this next function does.

```
sim<-function(type_distribution,para_distribution,vol,r,m,theta)
{

# This function samples returns under the risk neutral distribution
# type_distribution is the type of distribution used
# para_distribution contains the parameters driving the distribution
# vol contains the current volatility
# r contains the risk free rate
# m contains the risk premium
# theta contains the result obtained from solving the theta condition

if (type_distribution=="GH")
{
alpha=para_distribution[1]
beta=para_distribution[2]
delta=para_distribution[3]
mu=para_distribution[4]
lambda=para_distribution[5]

# Change in the parameter under the RN distribution
beta=beta+vol*theta
# In the GH case, we use the rgh() function taken from the fSeries
# package
result=rgh(1,alpha,beta,delta,mu,lambda)
}

if (type_distribution=="MN")
{
phi=para_distribution[1]
mu1=para_distribution[2]
sigma1=para_distribution[3]
mu2=para_distribution[4]
sigma2=para_distribution[5]

# Change in the parameter under the RN distribution
m1=(r+m+vol*mu1)*theta
s1=(theta^2)*(vol*sigma1)^2
s1=s1/2

m2=(r+m+vol*mu2)*theta
s2=(theta^2)*(vol*sigma2)^2
s2=s2/2

A=phi*exp(m1+s1)
B=phi*exp(m1+s1)+(1-phi)*exp(m2+s2)

# The new set of parameters is then:
phi=A/B
mu1=mu1+vol*(sigma1^2)*theta
```

```
mu2=mu2+vol*(sigma2^2)*theta

# Sampling of the MN innovation
u=runif(1)
if (u<=phi){
  result=rnorm(1,mu1,sigma1)
}else{
  result=rnorm(1,mu2,sigma2)
}
}

if (type_distribution=="Jumps")
{
lambda=para_distribution[1]
mu=para_distribution[2]
sigma=para_distribution[3]

# Change in the parameter under the RN distribution
mu_tilde=mu+theta*vol*sigma^2
A=mu*theta*vol
B=((vol*theta*sigma)^2)*(1/2)
lambda_tilde=lambda*exp(A+B)

# Sampling of the Jump innovation
z=rnorm(1)
N=rpois(1,lambda_tilde)
for (i in 0:N)
{
if (i>0)
{
  temp=rnorm(1,mu_tilde,sigma)
  z=z+temp
}
}

result=z
}

return(result)
}
```

Finally, the following function can be used to compute option prices, using the four previous ones. This option pricer also uses a variance reduction tool called the Duan and Simonato (1998)'s empirical martingale simulation method that is presented in Sect. 3.6. This empirical martingale simulation approach is a precious tool to reduce the number of simulations required to compute option prices. What is more (see Sect. 3.6), when used as a way to turn historical simulations into risk-neutral ones—as it is able to change the drift of the sampled returns—it delivers interesting performances that can even be better than those obtained with other pricing kernels, as presented in Chorro et al. (2010).

```
pricer<-function(option_spec,N,para_distribution,
para_vol,vol_start,lambda0,type_vol,type_distribution){

# This function computes option prices by Monte Carlo, using the
# functions "variance", "sim" and "solve_theta".

# option_spec is a list containing the option's parameters
```

```
# N is the number of simulations
# para_distribution is a vector containing the distribution's
# parameters
# para_vol is a vector containing the volatility's parameters
# vol_start is the volatility starting value
# lambda0 is the risk premium
# type_vol is the type of volatility structure
# type_distribution is the type of distribution

T=book_optim$T
S=book_optim$S
K=book_optim$K
r=book_optim$r/250

# T_max est exprimé en jours
T=T*250
T=round(T,0)

##########################
# Step 1: sampling the returns
base_sim=matrix(0,T,N)
for (nsim in 1:N)
{
  vol=vol_start
  vol_stock=c()
  innovation=0
  for (n_t in 1:T)
  {

  m=lambda*vol-(vol^2)/2
  h_t=vol^2
  theta=solve_theta(type_distribution,para_distribution,vol,r,m)
  innovation=sim(type_distribution,para_distribution,vol,r,m,theta)
  if (type_distribution=="Jumps")
  {
  innovation=innovation+theta*vol
  }
  vol=variance(type_vol,para_vol,vol,innovation)
  vol_stock=c( vol_stock,vol)
  val=r-(vol^2)/2+vol*innovation

  base_sim[n_t,nsim]=val
  }
}

##########################
# Step 2: turning returns into Monte-Carlo prices
x=base_sim
x=apply(x,2,cumsum)
x=exp(x)
x=S*x

##########################
# Step 3: martingalization of the sample
z=apply(x,1,mean)
z=S*exp(r*(1:T))/z
x=x*z
P=x

##########################
# Step 4: Computation of the option prices
```

```
P=P[nrow(P_temp),]
sim_temp=P_temp-K
sim_temp=pmax(sim_temp,0,na.rm=TRUE)
sim_temp=mean(sim_temp,na.rm=TRUE)
sim_temp=sim_temp*exp(-r*T)
result=sim_temp

return(result)
}
```

Tables 4.10 and 4.11 and Figs. 4.15, 4.16 and 4.17 present the pricing results obtained with the time series models estimated previously, jointly with the R functions detailed earlier. Several empirical conclusions can be reached from these tables and graphs:

- The graphs and the tables deliver a very encouraging message for this time series approach to option pricing, at least when it comes to the EGARCH-based models. Globally speaking, when combining non-Gaussian distributions and an EGARCH model, we obtain an additional reduction in the AARPE when compared to the Heston and Nandi (2000) case. This comparison is done in Table 4.11, under the column REC (for recursive estimation strategy): for example, in the EGARCH-MN case, the AARPE is equal to 21 % when in the HN one it was equal to 24 %. It is not a major improvement, but it still leads to a decrease in this criterion, highlighting the contribution of extreme distributions to option pricing.
- The second interesting conclusion—again in the case of EGARCH—is that for the cases considered here, the AARPE is similar or smaller than the one obtained in the case of the HN model with a quadratic pricing kernel: the AARPE is equal to 22 % in the GH case and to 20 % in the Jump case, when the HN model with a quadratic pricing kernel has an AARPE equal to 22 % over our experiments. Hence, the addition of a non-Gaussian distribution to a GARCH-based variance dynamics has somewhat a similar pricing impact to that of using more complex risk preferences. This is result is consistent with the findings in Guégan et al. (2013).
- Let us turn shortly to the APARCH case: across all our experiments, this model always delivers the worst performances. This seemingly come from the additional complexity in this model that is related to the estimation of the exponent in the variance dynamics Δ. Using our estimates, its pricing performances range between 38 and 51 %. Compared to the EGARCH model which is based on the modelling of the logarithm of the variance—that stands a good chance to be Gaussian, making things easier in terms of estimation[33]—the numerical difficulties coming from the estimation of the APARCH are a lot higher. Still, using the REC estimation strategy, the option pricing performance is a lot better as seen from Table 4.11: this table compares the pricing performances

[33]See e.g. Andersen et al. (2001).

Table 4.10 Empirical comparison of the pricing performances obtained across time series models

Time to maturity ≤ 0.3 year										
Model/Moneyness	0.80	0.84	0.89	0.93	0.98	1.02	1.07	1.11	1.16	1.20
HN Calibrated	1.24	1.48	2.00	3.01	6.05	8.90	10.66	11.36	18.92	34.41
HN Estimated	1.37	1.70	2.19	3.95	7.35	15.57	34.95	60.45	77.99	88.40
HN Estimated Quad.	1.33	1.68	2.60	4.22	8.09	17.86	46.71	97.23	108.34	115.31
EGARCH-GH	2.09	2.45	2.99	3.60	5.49	10.59	23.29	57.42	79.29	91.71
EGARCH-MN	2.13	2.42	2.83	3.24	5.12	10.36	23.36	50.54	67.20	79.09
EGARCH-Jumps	1.95	2.08	2.36	2.97	5.14	10.16	21.47	46.62	63.06	84.84
APARCH-GH	3.66	4.62	5.53	6.95	10.23	17.03	32.46	66.89	87.65	110.95
APARCH-MN	3.57	4.54	5.72	7.43	10.65	16.98	31.45	73.18	96.31	121.93
APARCH-Jumps	3.37	4.31	5.40	7.05	11.43	19.99	41.28	93.54	121.04	125.45
$0.3 \leq$ Time to maturity ≤ 0.5 year										
Model/Moneyness	0.80	0.84	0.89	0.93	0.98	1.02	1.07	1.11	1.16	1.20
HN Calibrated	2.44	2.57	3.17	4.38	6.69	8.69	8.58	6.46	8.82	19.56
HN Estimated	2.66	3.20	4.12	5.96	9.58	15.12	23.44	37.99	59.47	74.35
HN Estimated Quad.	2.57	3.56	4.54	5.97	7.88	10.36	15.83	22.65	31.81	56.77
EGARCH-GH	3.91	4.75	5.53	6.57	8.56	11.16	17.95	27.44	41.80	67.26
EGARCH-MN	3.81	4.89	5.44	6.71	9.11	11.70	17.14	26.44	37.77	54.37
EGARCH-Jumps	2.93	3.91	4.30	5.17	7.51	10.53	16.57	25.33	34.66	49.51
APARCH-GH	7.94	9.82	11.25	14.44	18.96	23.48	33.28	47.47	63.65	86.19
APARCH-MN	7.86	9.63	11.54	14.48	18.23	23.35	31.21	42.76	58.24	81.94
APARCH-Jumps	8.94	10.25	12.31	15.96	21.79	29.44	41.32	62.21	87.11	121.90
Time to maturity ≥ 0.5 year										
Model/Moneyness	0.80	0.84	0.89	0.93	0.98	1.02	1.07	1.11	1.16	1.20
HN Calibrated	3.88	3.60	3.53	4.10	4.97	5.32	5.23	4.72	5.51	16.92
HN Estimated	4.88	5.53	5.65	7.65	9.38	14.50	18.07	26.51	37.43	53.40
HN Estimated Quad.	5.29	6.47	7.00	7.83	9.19	10.20	12.40	14.02	18.86	27.65
EGARCH-GH	6.62	8.55	9.44	10.62	13.49	15.67	20.15	23.87	33.35	48.09
EGARCH-MN	6.93	8.73	9.86	11.04	14.23	16.76	21.64	26.59	34.85	50.56
EGARCH-Jumps	5.70	6.99	7.51	8.81	11.39	14.55	18.81	24.25	33.93	49.13
APARCH-GH	13.66	17.69	20.36	24.23	29.97	35.03	45.94	54.25	68.95	98.79
APARCH-MN	13.98	17.14	19.94	23.70	29.15	34.68	44.98	53.69	67.28	93.89
APARCH-Jumps	16.49	20.13	24.20	29.43	37.79	47.22	63.92	80.17	113.73	166.23

model by model depending on the estimation strategy used or depending on the selected pricing kernel. The first and third columns make it possible to compare the improvement coming from the recursive estimation method over the QML one in terms of pricing errors. In the case of EGARCH-based time series models, there are little changes between those two estimation strategies: this is consistent with the idea that the EGARCH model has a low numerical complexity. For the APARCH results, we obtain a different story: the AARPE obtained by QML is

Table 4.11 Pricing performances (from the AARPE criterion) obtained using different estimation strategies for each model's parameters

	Estimation strategy/Pricing Kernel				
	QML/Ess	QML/Quad-Ess	REC/Ess	REC/Mart	Calib/Esscher.
EGARCH-GH	0.19	–	0.22	0.19	–
EGARCH-MN	0.22	–	0.21	0.19	–
EGARCH-Jumps	0.21	–	0.20	0.17	–
APARCH-GH	0.62	–	0.38	0.34	–
APARCH-MN	0.74	–	0.38	0.35	–
APARCH-Jumps	0.79	–	0.51	0.45	–
HN	0.24	0.22	–	–	0.07

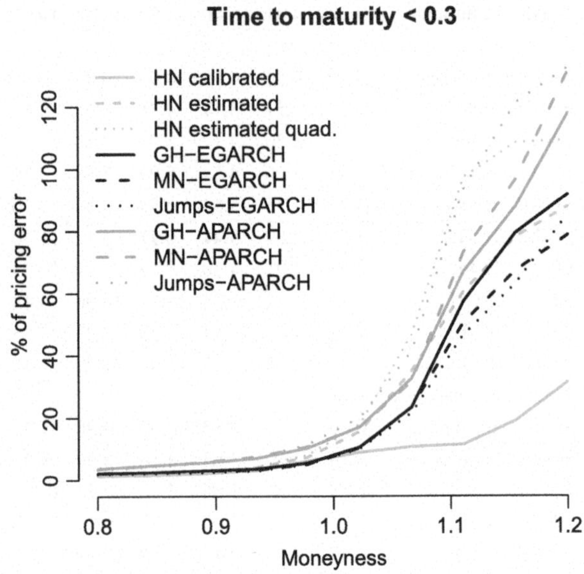

Fig. 4.15 Comparisons of the pricing performances across time series models, for options with a time to maturity below 0.3 year

almost consistently twice the one obtained using the REC estimates. Again, the APARCH structure makes it more difficult to obtain reliable estimates than in the EGARCH case, explaining part of the differences that we obtain when it comes to the pricing results.

- Turning back to the results obtained from the EGARCH model, Figs. 4.15, 4.16 and 4.17 allow the reader to understand where the pricing performance of those time series models is the best. The strongest improvement is obtained for options of a mid-range maturity: for options whose time-to-maturity lies between 3 and 6 months, the performances obtained with the EGARCH-based models are roughly comparable or better than those obtained with the Heston and Nandi (2000) model with a quadratic pricing kernel. Compared to the plain HN model, the improvement is very important for a moneyness around 1, for which the reduction in AARPE can be roughly up to 50 %. For options with a time-to-maturity below

Fig. 4.16 Comparisons of the pricing performances across time series models, for options with a time to maturity between 0.3 and 0.5 year

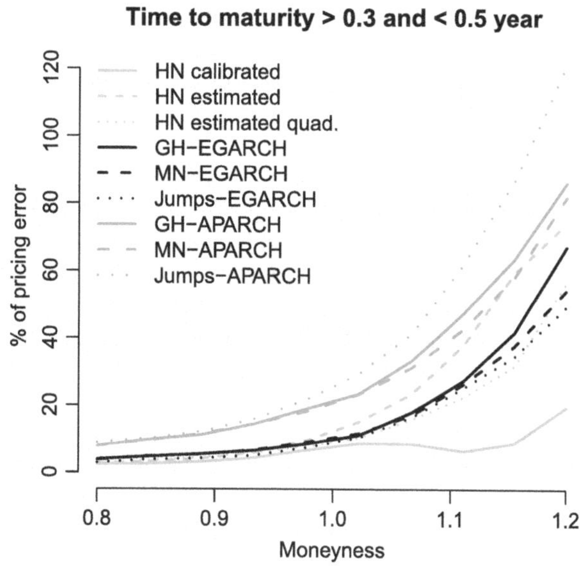

Fig. 4.17 Comparisons of the pricing performances across time series models, for options with a time to maturity higher than 0.5 year

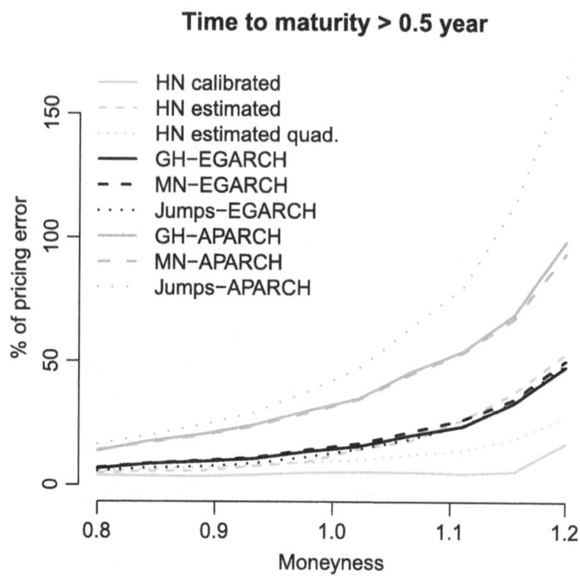

3 months, the times series models based on an EGARCH models deliver pricing performances that are either comparable or better than that of the Heston and Nandi (2000) model. Given the poorer performances of the HN model with a quadratic pricing kernel, the EGARCH models are doing a better pricing job. Finally, when it comes to options with a time to maturity larger than 6 months, the EGARCH models deliver a performance that is similar to that of the classic HN one, but poorer than that of the HN model based on a quadratic pricing kernel.

The fact that this latter approach dominates is certainly due to the fact that it uses the VIX in its pricing, therefore accounting for the actual volatility premium with a higher degree of precision than the EGARCH models for which this premium is model-induced.

As a last remark, Table 4.11 compares the results obtained using the REC estimates and an exponential affine pricing kernel to the results obtained by simply applying the EMS of Duan and Simonato (1998) under the column "REC-Mart".[34] The use of this variance reduction method somewhat as a pricing kernel (see Sect. 3.6) improves again the pricing results. For example, in the case of EGARCH-MN model, the AARPE drops by another 2 % points again. The best score is obtained for the EGARCH-Jump model, whose AARPE is equal to 17 %, 10 % points away from the calibration AARPE of 7 %. This is related to the fact that this change in the probability measure is somewhat not transmitted to the variance dynamics, unlike the exponential affine pricing kernel case. Given the simplicity of the approach and the consistent improvement over the results, this approach—combines with a non-Gaussian EGARCH-based model—looks as an interesting practical approach.

4.4 Conclusion

As a conclusion from these empirical experiments, the EGARCH-based models seem to offer an interesting contribution to the pricing of options: by incorporating non-Gaussian distributions, the pricing results are increasingly closer to that obtained from the calibrated HN model. It should be seen as an alternative to the use of a quadratic pricing kernel: instead of improving the specification of preferences towards risk, a better modeling of the statistical mechanics of financial markets is therefore a legitimate approach to improve option pricers. The interest of this exercise—beyond the understanding of how to use time series analysis to price options—is probably the perspectives opened by the empirical results obtained in the cited literature and in our tables and graphs: this time series approach to option pricing can be successfully applied to obtain prices that are rather close to those actually priced by financial markets. This also means that such an approach can be used to approximate the price of options for markets that are either illiquid or that have no existing option markets. This field of research is rather new: there will be many more research articles on this question, and the extent to which time series analysis can actually be used to price contingent claims will be gauged through their results. For now, it remains a vivid research stream that looks from our results' perspectives rather promising.

[34]Those results are still based on the use of the REC estimates.

References

Ait-Sahalia Y, Lo A (1998) Nonparametric estimation of state-price densities implicit in financial asset prices. J Financ 53:499–547

Amisano G, Giacomini R (2007) Comparing density forecasts via weighted likelihood ratio tests. J Bus Econ Stat Am Stat Assoc 25:177–190

Andersen TG, Bollerslev T, Diebold FX Labys P (2001) The distribution of realized exchange rate volatility. J Am Stat Assoc 96:42–55

Badescu A, Kulperger R (2008) GARCH option pricing: a semiparametric approach. Insur Math Econ 43(1):69–84

Badescu A, Elliott RJ, Kulperger R, Miettinen J, Siu TK (2011) A comparison of pricing kernels for GARCH option pricing with generalized hyperbolic distributions. Int J Theor Appl Financ 14(5):669–708

Bai J, Ng S (2002) Determining the number of factors in approximate factor models. Econometrica 70:191–221

Barone-Adesi G, Engle R, Mancini L (2008) A GARCH option pricing model with filtered historical simulation. Rev Financ Stud 21:1223–1258

Bates D (1988) Pricing options under jump-diffusion processes. Working Paper

Bates D (1996) Jumps and stochastic volatility: exchange rate processes implicit in deutsche mark options. Rev Financ Stud 9(1):69–107

Bates D (1997) The Skewness premium: option pricing under asymmetric processes. Adv Futures Options Res 9:51–82

Bertholon H, Monfort A, Pegoraro F (2008) Econometric Asset pricing modelling. J Financ Econ 6(4):407–458

Black F, Scholes M (1973) The pricing of options and corporate liabilities. J Polit Econ 81:637–659

Carr P, Madan D (1999) Option valuation using the fast fourier transform. J Comput Financ 2(4):61–73

Chacko G, Viceira LM (2003) Spectral GMM estimation of continuous-time processes. J Econ 116(1–2):259–292

Chamberlain G, Rothschild M (1983) Arbitrage, factor structure, and mean variance analysis on large asset markets. Econometrica 51:1281–1304

Chorro C, Guégan D, Ielpo F (2010) Martingalized historical approach for option pricing. Financ Res Lett 7(1):24–28

Chorro C, Guégan D, Ielpo F, Lalaharison H (2014) Testing for leverage effect in financial returns. CES Working Papers, 2014.22

Christoffersen P, Heston SL, Jacobs K (2006) Option valuation with conditional skewness. J Econ 131:253–284

Cody WJ (1969) Rational Chebyshev approximations for the error function. Math Comput 23:631–637

Cont R, da Fonseca J (2002) Dynamics of implied volatility surfaces. Quant Finan 2(1):45–60

Cooley JW, Tukey JW (1965) An algorithm for the machine calculation of complex Fourier series. Math Comput 19:297–301

Duan JC, Simonato JG (1998) Empirical martingale simulation for asset prices. Manag Sci 44:1218–1233

Duffie D, Kan R (1996) A yield-factor model of interest rates. Math Financ 6:379–406

Guégan D, Ielpo F, Lalaharison H (2013) Option pricing with discrete time jump processes. J Econ Dyn Control 37(12):2417–2445

Heston S (1993) A closed-form solution for options with stochastic volatility with applications to bond and currency options. Rev Financ Stud 6(2):27–343

Heston SL, Nandi S (2000) A closed-form garch option valuation. Rev Financ Stud 13:585–625

Jiang GJ, Knight JL (2002) Estimation of continuous-time processes via the empirical characteristic function. J Bus Econ Stat 20(2):198–212

Jiang GJ, Tian Y (2005) The model-free implied volatility and its information content. Rev Financ Stud 18(4):1305–1342

Rockinger M, Semenova M (2005) Estimation of jump-diffusion process via empirical characteristic function. FAME Research Paper Series rp150, International Center for Financial Asset Management and Engineering.

Shimko D (1993) Bounds of probability. RISK 6:33–37

Vuong QH (1989) Likelihood ratio tests for model selection and non-nested hypotheses. Econometrica 57:307–333

Walker JS (1996) Fast Fourier transforms, 2nd edn. CRC Press, Boca Raton

Mathematical Appendix

We present in this concise appendix the main probabilistic notions used in the previous chapters. For more details, the diligent reader may refer to Jacod and Protter (2004) for the classical Probability theory, to Williams (1991) for martingales and to Lapeyre et al. (2013) and Robert and Casella (2010) for Monte Carlo methods.

Let $(\Omega, \mathcal{A}, P)$ be a probability space where Ω is a set, $\mathcal{A}$ a σ-algebra and P a probability defined on $\mathcal{A}$. In this section we denote by E the expectation under the probability P. For $p \in \mathbb{N}^*$, we classically denote by $L^p(\Omega, \mathcal{A}, P)$ the set of the real random variables $Z : (\Omega, \mathcal{A}) \to \mathbb{R}$ such that $E[|Z|^p] < \infty$.

Gaussian Random Variables

A real random variable X follows a Gaussian distribution of mean m and variance σ^2 (the standard notation is $\mathcal{N}(m, \sigma^2)$) if its density function is equal to

$$\frac{1}{\sqrt{2\pi\sigma^2}} e^{-\frac{(x-m)^2}{2\sigma^2}}.$$

The associated moment generating function is given by

$$E[e^{tX}] = e^{tm+\frac{\sigma^2 t^2}{2}}, \ \forall t \in \mathbb{R}.$$

There are no explicit expressions for the distribution function

$$N(x) = P(X \leq x)$$

of a $\mathcal{N}(m, \sigma^2)$, but due to the importance of this quantity in mathematical finance (in particular in the Black and Scholes formula (4.3)) we mention that very accurate approximations for N and N^{-1} exist (see Cody 1969) and are implemented in R via the *pnorm* and *qnorm* functions of the *stats* package.

The following result gives a simple method (known as the Box-Muller method) to generate samples of Gaussian distributions from classical random number

C. Chorro et al., *A Time Series Approach to Option Pricing*,
DOI 10.1007/978-3-662-45037-6

generators. In R, this method may be used via the *rnorm* command selecting *RNGkind(normal.kind =Box-Muller)* in the *stats* package.

Proposition A.1 *If U_1 and U_2 are two independent uniform random variables on* $[0, 1]$ *then*

$$G_1 = \sqrt{-2log(U_1)}cos(2\pi U_2) \text{ and } G_2 = \sqrt{-2log(U_1)}sin(2\pi U_2)$$

are two independent $\mathcal{N}(0, 1)$.

Proof Let us define

$$\Phi(x, y) = (u = \sqrt{-2log(x)}cos(2\pi y), v = \sqrt{-2log(x)}sin(2\pi y)).$$

It is easy to see that Φ is a C^1-diffeomorphism[1] from $]0, 1[^2$ into $\mathbb{R}^2-(\mathbb{R}_+\times\{0\})$ with a Jacobian determinant fulfilling $|J(\Phi)(x, y)| = \frac{2\pi}{x}$. Since $u^2+v^2 = -2log(x)$, we have $|J(\Phi^{-1})(u, v)| = \frac{1}{2\pi}e^{-\frac{u^2+v^2}{2}}$. According to the change of variables theorem, for any $F : \mathbb{R}^2 \to \mathbb{R}$ continuous and bounded

$$\int_{]0,1[^2} F(\Phi(x, y))dxdy = \int_{\mathbb{R}^2-(\mathbb{R}_+\times\{0\})} F(u, v)\frac{1}{2\pi}e^{-\frac{u^2+v^2}{2}} dudv$$

and the result follows. □

Conditional Expectation

Let X be a random variable in $L^1(\Omega, \mathcal{A}, P)$ and let $\mathcal{G}$ denote a sub σ-algebra of $\mathcal{A}$.

Definition A.1 There exists a random variable $Z \in L^1(\Omega, \mathcal{A}, P)$ such that

$$i)\ Z \text{ is } \mathcal{G}\text{-measurable}$$

$$ii)\ E[XU] = E[ZU],\ \forall U\ \mathcal{G}\text{-measurable and bounded.}$$

[1] If $\mathcal{O}_1$ and $\mathcal{O}_2$ are two open sets of $\mathbb{R}^n$, we say that $\Phi : \mathcal{O}_1 \to \mathcal{O}_2$ is a C^1-diffeomorphism if it is a bijection that is C^1 and if Φ^{-1} is C^1 as well. In this case, we define by $J(\Phi)$ the determinant of the matrix of the partial derivatives of Φ called the Jacobian and we have the change of variables theorem: $\forall F : \mathcal{O}_2 \to \mathbb{R}$ continuous and bounded,

$$\int_{\mathcal{O}_2} F(y)dy = \int_{\mathcal{O}_1} F(\Phi(x))|J(\Phi)(x)|dx.$$

Z is denoted by $E[X|\mathcal{G}]$ and is called the conditional expectation of X given $\mathcal{G}$. Moreover, Z is unique up to almost-sure equality.

Remark A.1 When Y is a random variable, $E[X|Y]$ is simply defined as $E[X|\sigma(Y)]$ where $\sigma(Y)$ is the smallest σ-algebra that makes Y measurable. In particular, when the pair (X, Y) owns a density $f_{(X,Y)}$ with respect to the Lebesgue measure on $\mathbb{R}^2$, the marginal densities of X and Y are given by

$$f_X(x) = \int_{\mathbb{R}} f_{(X,Y)}(x, y)dy \text{ and } f_Y(y) = \int_{\mathbb{R}} f_{(X,Y)}(x, y)dx.$$

If X and Y are independent, we have $f_{(X,Y)} = f_X f_Y$ and if the condition of independence is relaxed, we obtain the following disintegration formula:

$$f_{(X,Y)}(x, y) = f_{X|Y}(x, y) f_Y(y)$$

where

$$f_{X|Y}(x, y) = \frac{f_{(X,Y)}(x, y)}{f_Y(y)}$$

if $f_Y(y) \neq 0$ and $f_{X|Y}(x, y) = 0$ otherwise. The function $f_{X|Y}$ is called the conditional density of X given Y. In fact, if ϕ satisfies $\phi(X) \in L^1(\Omega, \mathcal{A}, P)$,

$$E[\phi(X)|Y] = \Phi(Y) \text{ with } \Phi(y) = \int_{\mathbb{R}} \phi(x) f_{X|Y}(x, y)dx.$$

The practical computations involving conditional expectations are simplified by the following properties that are used all along this book:

Proposition A.2 *Let X and Y be two random variables in $L^1(\Omega, \mathcal{A}, P)$ and let $\mathcal{G}$ be a sub σ-algebra of $\mathcal{A}$, then,*

(a) (Positivity) If $X \geq 0$ P-a.s, $E[X|\mathcal{G}] \geq 0$ P-a.s.
(b) (Linearity) If $(\alpha, \beta) \in \mathbb{R}^2$, $E[\alpha X + \beta Y|\mathcal{G}] = \alpha E[X|\mathcal{G}] + \beta E[Y|\mathcal{G}]$ P-a.s.
(c) (Monotony) If $X \geq Y$ P-a.s, $E[X|\mathcal{G}] \geq E[Y|\mathcal{G}]$ P-a.s.
(d) (Jensen inequality) If $\psi : \mathbb{R} \to \mathbb{R}$ is a convex function such that $\psi(X) \in L^1(\Omega, \mathcal{A}, P)$, then, $\psi(E[X|\mathcal{G}]) \leq E[\psi(X)|\mathcal{G}]$ P-a.s.
(e) $E[E[X|\mathcal{G}]] = E[X]$.
(f) If X is $\mathcal{G}$-measurable, $E[X|\mathcal{G}] = X$ P-a.s.
(g) (Taking out what is known) If Y is $\mathcal{G}$-measurable and bounded, $E[XY|\mathcal{G}] = YE[X|\mathcal{G}]$ P-a.s.
(h) (Role of independence) If X is independent of $\mathcal{G}$, $E[X|\mathcal{G}] = E[X]$ P-a.s.
(i) (Tower property) If $\mathcal{G}'$ is a sub σ-algebra of $\mathcal{A}$ such that $\mathcal{G}' \subset \mathcal{G}$, then, $E[E[X|\mathcal{G}]|\mathcal{G}'] = E[X|\mathcal{G}']$ P-a.s.

The preceding properties are often sufficient to compute conditional expectations and we only come back to the definition for more difficult cases. In particular, the following result is useful for the study of financial models:

Proposition A.3 *If X is independent of $\mathcal{G}$ and if Y is $\mathcal{G}$-measurable, then, for all measurable function $\Phi : \mathbb{R}^2 \to \mathbb{R}$ such that $E[|\Phi(X, Y)|] < \infty$, we have*

$$E[\Phi(X, Y)|\mathcal{G}] = \psi(Y) \text{ where } \psi(y) = E[\Phi(X, y)].$$

Proof We have $\psi(y) = \int_{\mathbb{R}} \Phi(x, y)dP_X(x)$ and the measurability of ψ is a classical consequence of Fubini's theorem. For $G \in \mathcal{G}$, we set $Z = 1_G$. We deduce from the hypotheses that $P_{(X,Y,Z)} = P_X \otimes P_{(Y,Z)}$, thus,

$$E[\Phi(X, Y)1_G] = \int_{\mathbb{R}} \int_{\mathbb{R}^2} \Phi(x, y)zdP_{(Y,Z)}(y, z)dP_X(x).$$

By Fubini's theorem,

$$E[\Phi(X, Y)1_G] = \int_{\mathbb{R}^2} \psi(y)zdP_{(Y,Z)}(y, z)$$

so

$$E[\Phi(X, Y)1_G] = E[\psi(Y)1_G]$$

which completes the proof. □

The notion of filtration is used in mathematical finance to represent the evolution of financial information along time. It is presented below in its discrete time version:

Definition A.2

(a) A non decreasing family $(\mathcal{F}_t)_{t\in\{0,\dots,T\}}$ of sub σ-algebras of $\mathcal{A}$ is called a filtration. This filtration is said to be complete when it contains negligible sets that is $\forall t \in \{0, \dots, T\}, \mathcal{N} \subset \mathcal{F}_t$ where

$$\mathcal{N} = \{N \subset \Omega; \exists A \in \mathcal{A}, N \subset A, P(A) = 0\}.$$

(b) A family of random variables $(X_t)_{t\in\{0,1,\dots,T\}}$ is adapted to the filtration $(\mathcal{F}_t)_{t\in\{0,\dots,T\}}$ if $\forall t \in \{0, \dots, T\}$, X_t is $\mathcal{F}_t$ measurable.
(c) A family of random variables $(X_t)_{t\in\{0,\dots,T\}}$ is predictable with respect to the filtration $(\mathcal{F}_t)_{t\in\{0,1,\dots,T\}}$ if $\forall t \in \{0, \dots, T-1\}$, X_{t+1} is $\mathcal{F}_t$ measurable and if X_0 is $\mathcal{F}_0$ measurable.

We have seen in Sect. 3.2 that the absence of arbitrage opportunities in financial models is intrinsically linked to the notion of martingales that is defined below:

Definition A.3 A family of random variables $(M_t)_{t\in\{0,\dots,T\}}$ in $L^1(\Omega, \mathcal{A}, P)$ that is adapted to the filtration $(\mathcal{F}_t)_{t\in\{0,\dots,T\}}$ is called:

- a martingale if for $T \geq t \geq s \geq 0$, $E[M_t|\mathcal{F}_s] = M_s$.
- a supermartingale if for $T \geq t \geq s \geq 0$, $E[M_t|\mathcal{F}_s] \leq M_s$.
- a submartingale if for $T \geq t \geq s \geq 0$, $E[M_t|\mathcal{F}_s] \geq M_s$.

The following proposition is the key stone of the definition of the stochastic discount factor associated to an equivalent martingale measure (see Sect. 3.2.2):

Proposition A.4 *Let* $(M_t)_{t\in\{0,\dots,T\}}$ *be a non negative martingale such that* $E[M_T] = 1$. *If* X *is a random variable in* $L^1(\Omega, \mathcal{A}, P)$ *that is* $\mathcal{F}_t$ *measurable, then, for* $t \geq s \geq 0$

$$E_Q[X|\mathcal{F}_s] = \frac{1}{M_s} E[XM_t|\mathcal{F}_s]$$

where Q *is the probability defined by the density function* $\frac{dQ}{dP} = M_T$ *with respect to* P.

Proof First, when $s \leq T$, for an $\mathcal{F}_s$ measurable and bounded random variable Z we may deduce from the martingale property of (M_t) that

$$E_Q[Z] = E[ZM_T] = E[M_s Z].$$

Moreover, for $t \geq s \geq 0$, we have from Proposition A.2

$$E_Q[XZ] = E[M_T XZ] = E[E[XM_T|\mathcal{F}_s]Z] = E[E[XE[M_T|\mathcal{F}_t]|\mathcal{F}_s]Z]$$

thus

$$E_Q[XZ] = E[E[M_t X|\mathcal{F}_s]Z] = E\left[\frac{M_s}{M_s} E[XM_t|\mathcal{F}_s]Z\right] = E_Q\left[\frac{1}{M_s} E[XM_t|\mathcal{F}_s]Z\right].$$

Hence, from Definition A.1,

$$E_Q[X|\mathcal{F}_s] = \frac{1}{M_s} E[XM_t|\mathcal{F}_s].$$

□

Monte Carlo Methods

In this section, $\underset{a.s}{\to}$, $\underset{L^1}{\to}$ and $\underset{\mathcal{D}}{\to}$ denote the almost sure convergence, the convergence in $L^1(\Omega, \mathcal{A}, P)$ and the convergence in distribution for sequences of random variables. Theoretical foundations of Monte Carlo Methods are mainly based on two fundamental asymptotic results: The Strong Law of Large Numbers and the Central Limit Theorem. The Strong Law of Large Numbers ensures that, under integrability conditions, the mean of a sequence of i.i.d random variables is an approximation of the expectation:

Theorem A.1 *Let (X_n) be a sequence of i.i.d random variables.*

(a) Suppose that $X \in L^1(\Omega, \mathcal{A}, P)$. Denoting $S_n = X_1 + \ldots + X_n$, we have

$$\frac{S_n}{n} \underset{a.s \text{ and } L^1}{\to} E[X_1].$$

(b) If $E[|X_1|] = +\infty$, the sequence S_n diverges almost surely.

The Central Limit Theorem gives some precisions concerning the speed of convergence in the Strong Law of Large Numbers:

Theorem A.2 *Let (X_n) be a sequence of i.i.d random variables. Suppose that $X_1 \in L^2(\Omega, \mathcal{A}, P)$, then,*

$$\frac{S_{n-nm}}{\sqrt{n}\sigma} \underset{\mathcal{D}}{\to} \mathcal{N}(0,1)$$

where $m = E[X_1]$ and $\sigma^2 = Var[X_1]$.

The preceding theorem is classically used to build asymptotic confidence intervals. In fact, we obtain $\forall a \in \mathbb{R}_+$,

$$P\left(\frac{-a\sigma}{\sqrt{n}} \leq \frac{S_n}{n} - E[X_1] \leq \frac{a\sigma}{\sqrt{n}}\right) \underset{n\to\infty}{\to} \int_{-a}^{a} \frac{1}{\sqrt{2\pi}} e^{-\frac{x^2}{2}} dx.$$

In practice we know from tables that

$$P(|\mathcal{N}(0,1)| \leq 1.96) = 0.95$$

thus when n is large enough, with a confidence of 95 %,

$$E[X_1] \in \left[\frac{S_n}{n} - \frac{1.96\sigma}{\sqrt{n}}, \frac{S_n}{n} + \frac{1.96\sigma}{\sqrt{n}}\right].$$

The magnitude of the error is given by $\frac{3.92\sigma}{\sqrt{n}}$: the magnitude of n and σ is fundamental to control the length of the confidence interval. When σ is unknown, the so-called empirical variance gives us an unbiased and consistent estimator:

Proposition A.5 *Let $(X_n)_{n\in\mathbb{N}}$ be i.i.d random variables in $L^2(\Omega, \mathcal{A}, P)$. Then, if we define*

$$\hat{\sigma_n}^2 = \frac{n}{n-1}\left(\frac{1}{n}\sum_{i=1}^{n} X_i^2 - (\frac{1}{n}\sum_{i=1}^{n} X_i)^2\right)$$

we have $E[\hat{\sigma_n}^2] = \sigma^2$ and $\hat{\sigma_n}^2 \underset{a.s}{\rightarrow} \sigma^2$.

According to the following result, that is a direct consequence of the Slutsky's lemma,[2] the preceding estimator is used in practice to deduce confidence intervals from observations:

Proposition A.6 *Let $(X_n)_{n\in\mathbb{N}}$ be i.i.d random variables in $L^2(\Omega, \mathcal{A}, P)$, then,*

$$\frac{S_n - nE[X_1]}{\sqrt{n}\hat{\sigma_n}} \underset{\mathcal{D}}{\rightarrow} \mathcal{N}(0, 1).$$

Basically, to approximate the quantity $E[f(X)]$ by Monte Carlo Methods we have to

- Generate a n-sample of the distribution of X,
- Compute $\frac{1}{n}\sum_{k=1}^{n} f(X_k)$ for large n,
- Compute the confidence interval $\left[\frac{S_n}{n} - \frac{1.96\hat{\sigma_n}}{\sqrt{n}}, \frac{S_n}{n} + \frac{1.96\hat{\sigma_n}}{\sqrt{n}}\right]$ coming from the Central Limit Theorem.

This method is easy to implement on any software once we are able to generate samples of particular distributions. Moreover, only integrability conditions are required for f. Nevertheless, it is important to keep in mind that the precision of the method (measured by the size of the confidence interval) is a random variable depending on the magnitude of σ. From Proposition A.1, we have seen how to generate independent Gaussian random variables from uniform ones (that may

[2]If (X_n) and (Y_n) are two sequences of random variables such that $X_n \underset{\mathcal{D}}{\rightarrow} X$ and $Y_n \underset{\text{a.s}}{\rightarrow} c$ where c is a constant, then, $X_nY_n \underset{\mathcal{D}}{\rightarrow} Xc$.

be obtained using R with the *runif* command). We generalize this idea to the distributions involved in the empirical Chap. 4 namely the Poisson, the Mixture of two Gaussians and the Generalized Hyperbolic distributions:

Poisson distribution: A Poisson random variable X of parameter $\lambda \in \mathbb{R}_+^*$ is a random variable with values in $\mathbb{N}$ such that $\forall k \in \mathbb{N}$,

$$P(X = k) = p_k = \frac{e^{-\lambda}\lambda^k}{k!}.$$

As all discrete random variables, the Poisson distribution may be generated by the inversion method that is based on the following result:

Proposition A.7 *Let X be a random variable and F_X its distribution function. If U follows a uniform distribution on $[0, 1]$, then, $F_X^-(U)$ and X have the same distribution where F_X^- is the generalized inverse of F_X given $\forall u \in]0, 1[$ by*

$$F_X^-(u) = inf\{x \mid F_X(x) \geq u\}.$$

The preceding proposition is based on the elementary relation $F_X^-(u) \leq x \Leftrightarrow u \leq F_X(x)$ and in the case of the Poisson distribution

$$F_X^-(u) = \sum_{k=1}^{\infty} k \mathbf{1}_{\left\{\sum\limits_{j=0}^{k-1} p_j < u \leq \sum\limits_{j=0}^{k} p_j\right\}}.$$

For $\lambda \leq 10$ this method is implemented in R via the command *rpois* of the *stats* package. For $\lambda > 10$, this command uses the more complex and efficient methodology proposed in Ahrens and Dieter (1982).

Mixture of two Gaussian distributions: Generate mixture of two Gaussian distributions is an easy task remarking that

$$\mathbf{1}_{U \leq \phi} G_1 + \mathbf{1}_{U > \phi} G_2 \hookrightarrow MN(\phi, \mu_1, \mu_2, \sigma_1, \sigma_2)$$

when U, G_1 and G_2 are independent random variables such that U follows a uniform distribution on $[0, 1]$, $G_1 \hookrightarrow \mathcal{N}(\mu_1, \sigma_1^2)$ and $G_2 \hookrightarrow \mathcal{N}(\mu_2, \sigma_2^2)$.

Generalized Hyperbolic distributions: In Barndorff-Nielsen (1977) we find the normal variance-mean mixture representation of the $GH(\lambda, \alpha, \beta, \delta, \mu)$ distribution: We say that W follows an Generalized Inverse Gaussian distribution of parameter $(\lambda, \chi, \psi) \in \mathbb{R} \times (\mathbb{R}_+^*)^2$ (and we use the notation $W \hookrightarrow GIG(\lambda, \chi, \psi)$) if its density function is given $\forall x \in \mathbb{R}_+^*$ by

$$\frac{\left(\frac{\psi}{\chi}\right)^{\frac{\lambda}{2}}}{2K_\lambda(\sqrt{\psi\chi})} x^{\lambda-1} e^{-\frac{1}{2}(\chi x^{-1} + \psi x)}$$

where K_λ is the Bessel function of the third kind. If W and Z are two independent random variables such that $Z \hookrightarrow \mathcal{N}(0,1)$ and $W \hookrightarrow GIG(\lambda, \delta^2, \alpha^2 - \delta^2)$, then,

$$X = \mu + W\beta + \sqrt{W}Z \hookrightarrow GH(\lambda, \alpha, \beta, \delta, \mu).$$

Thus, we deduce from the preceding relation that GH distributions may easily be generated from *GIG* ones. In Dagpunar (1989), such an algorithm, based on the rejection method, is proposed for the *GIG* distribution and its numerical performances are discussed in Hörmann and Leydold (2014) where the authors prove its efficiency for $|\lambda| > 1$ and $\sqrt{\delta^2(\alpha^2 - \delta^2)} > 0.5$ (these conditions are compatible with the estimated parameters provided in Table 4.7). This method is implemented in R via the *rgh* command of the *fBasics* package.

Convergence of Discrete Time Markov Processes to Diffusions

We present here the basic techniques developed in Stroock and Varadhan (1979) that are classically applied to the study of the convergence of stochastic difference equations to diffusions.

For $T \in \mathbb{R}_+^*$ and $n \in \mathbb{N}^*$, we consider a Markov chain[3] indexed by the time unit $\tau = \frac{1}{n}$, $Z^{(n)} = (Y_{k\tau}^{(n)}, h_{k\tau}^{(n)})_{k \in \{0,\dots,nT\}}$, with values in $\mathbb{R}^2$ and starting from $(y_0, h_0) \in \mathbb{R}^2$. Then, $Z^{(n)}$ is embedded into a continuous time process $(Z_t^{(n)})_{t\in[0,T]}$ by defining

$$Z_t^{(n)} = Z_{k\tau}^{(n)} \text{ if } k\tau \le t < (k+1)\tau. \tag{A.1}$$

The sample paths of the latter process are by construction right continuous with left limit (cadlag). The next theorem gives general conditions to ensure the weak convergence[4] of $(Z_t^{(n)})_{t\in[0,T]}$ toward a bivariate diffusion.

Theorem A.3 *Let $p \in \mathbb{N}^*$ and μ (resp. Σ) be a continuous function from $\mathbb{R}^2$ into $\mathbb{R}^2$ (resp. into the set of the real matrix of size $2 \times p$). Suppose that for all $(r_1, r_2) \in (\mathbb{R}_+^*)^2$ and for some $s > 0$,*

[3]Let $(\Omega, \mathcal{A}, P)$ be a probability space equipped with a filtration $(\mathcal{F}_t)_{t\in\mathcal{T}}$, for some (totally ordered) index set $\mathcal{T}$ and let $(S, \mathcal{S})$ be a measurable space. An S-valued and adapted stochastic process $(X_t)_{t\in\mathcal{T}}$ is said to possess the Markov property with respect to the filtration $(\mathcal{F}_t)_{t\in\mathcal{T}}$ if, for each $A \in \mathcal{S}$ and each $s, t \in \mathcal{T}$ with $s < t$, $P(X_t \in A|\mathcal{F}_s) = P(X_t \in A|X_s)$. A Markov process is a stochastic process which satisfies the Markov property with respect to its natural filtration $(\mathcal{F}_t = \sigma(X_u; u \le t))$. When $\mathcal{T}$ is a discrete space, we say that $(X_t)_{t\in\mathcal{T}}$ is a Markov chain.

[4]Here, by weak convergence we mean weak convergence in the Skorokhod space of cadlag functions with values in $\mathbb{R}^2$ endowed with the Skorokhod topology (see e.g. Jacod and Shiryaev 2003, Section 6).

$$\lim_{\tau \to 0} \sup_{|y| \le r_1, |h| \le r_2} \left| \frac{1}{\tau} E\left[Z^{(n)}_{(k+1)\tau} - Z^{(n)}_{k\tau} | Z^{(n)}_{k\tau} = (y,h) \right] - \mu(y,h) \right| = 0, \tag{A.2}$$

$$\lim_{\tau \to 0} \sup_{|y| \le r_1, |h| \le r_2} \left| \frac{1}{\tau} Var\left[Z^{(n)}_{(k+1)\tau} - Z^{(n)}_{k\tau} | Z^{(n)}_{k\tau} = (y,h) \right] - \Sigma(y,h)\Sigma^t(y,h) \right| = 0, \tag{A.3}$$

$$\limsup_{\tau \to 0} \sup_{|y| \le r_1, |h| \le r_2} \tau^{\frac{-(2+s)}{2}} E\left[\left\| Z^{(n)}_{(k+1)\tau} - Z^{(n)}_{k\tau} \right\|^{2+s} | Z^{(n)}_{k\tau} = (y,h) \right] < \infty. \tag{A.4}$$

Then, if the stochastic differential equation

$$dZ_t = \mu(Z_t) + \Sigma(Z_t)dW_t, Z_0 = (y_0, h_0) \tag{A.5}$$

(where W is a p-dimensional standard Brownian motion) admits a unique weak solution[5] *on $[0,T]$, then the process $(Z^{(n)}_t)_{t\in[0,T]}$ weakly converges toward $(Z_t)_{t\in[0,T]}$ when τ goes to zero.*

From Moment Generating Functions to Option Prices

Here, using the notations of Sect. 3.7, we briefly remind how to obtain, up to numerical integration, option prices from the moment generating function of the logarithm of the risky asset.

Let us denote by $d_{t,T}$ the density function, under an EMM $\mathbb{Q}$, of $log(S_T)$ given $\mathcal{F}_t$. Thus, the arbitrage-free price, at time t, of a European call option with strike K and maturity T is given by

$$e^{-r(T-t)} E_{\mathbb{Q}}[(S_T - K)_+ \mid \mathcal{F}_t] = e^{-r(T-t)} \int_{log(K)}^{+\infty} (e^{\phi} - K) d_{t,T}(\phi) d\phi.$$

Exploiting the fact that $d_{t,T}$ and

$$d^*_{t,T} = \frac{e^x d_{t,T}}{\mathbb{G}^{\mathbb{Q}}_{log(S_T)|\mathcal{F}_t}(1)} = \frac{e^x d_{t,T}}{e^{r(T-t)} S_t}$$

are densities of probability, we use twice the classical inversion formula (see Stuart and Ord 1994, Chapter 4)

$$\int_{log(K)}^{+\infty} d(\phi) d\phi = \frac{1}{2} + \frac{1}{\pi} \int_0^{+\infty} \mathcal{R}e\left[\frac{K^{-i\phi}\mathbb{G}(i\phi)}{i\phi} \right] d\phi \tag{A.6}$$

[5]See for example Nelson (1990) for classical conditions.

(where d is a density function and $\mathbb{G}$ the associated moment generating function) to obtain

$$e^{-r(T-t)} E_{\mathbb{Q}}[(S_T - K)_+ \mid \mathcal{F}_t] = \frac{S_t}{2} + \frac{e^{-r(T-t)}}{\pi} \int_0^{+\infty} \mathcal{R}e\left[\frac{K^{-i\phi}\mathbb{G}^{\mathbb{Q}}_{log(S_T)|\mathcal{F}_t}(i\phi+1)}{i\phi}\right] d\phi - Ke^{-r(T-t)}\Big(\frac{1}{2} + \frac{1}{\pi} \int_0^{+\infty} \mathcal{R}e\left[\frac{K^{-i\phi}\mathbb{G}^{\mathbb{Q}}_{log(S_T)|\mathcal{F}_t}(i\phi)}{i\phi}\right] d\phi\Big).$$

References

Ahrens JH, Dieter U (1982) Computer generation of poisson deviates from modified normal distributions. ACM Trans Math Softw 8:163–179

Barndorff-Nielsen OE (1977) Exponentially decreasing distributions for the logarithm of particle size. Proc R Soc Lond Ser A 353:401–419

Cody WJ (1969) Rational chebyshev approximations for the error function. Math Comput 23:631–637

Dagpunar JS (1989) An easily implemented generalized inverse gaussian generator. Commun Stat B Simul Comput 18:703–710

Hörmann W, Leydold J (2014) Generating generalized inverse gaussian random variates. Stat Comput 24(4):547–557

Jacod J, Protter P (2004) Probability essentials. Springer, Berlin

Jacod J, Shiryaev AN (2003) Limit theorems for stochastic processes, 2nd edn. Springer, Berlin

Lapeyre B, Sulem A, Talay D (2013) Understanding numerical analysis for financial models. Cambridge University Press, Cambridge

Nelson DB (1990) ARCH models as diffusion approximations. J Econ 45:7–38

Robert CP, Casella G (2010) Introducing monte carlo methods with R. Springer, New York

Stroock DW, Varadhan SRS (1979) Multidimensional diffusion processes. Springer, New York

Stuart A, Ord JK (1994) Kendall's advanced theory of statistics. Vol. 1: Distribution theory, 6th edn. Wiley-Blackwell, New York

Williams D (1991) Probability with martingales. Cambridge University Press, Cambridge

Index

C. Chorro et al., *A Time Series Approach to Option Pricing*,
DOI 10.1007/978-3-662-45037-6